GNU Astronomy Utilities

A catalogue record for this book is available from the Hong Kong Public Libraries.

Published in Hong Kong by Samurai Media Limited.

Email: info@samuraimedia.org

ISBN 978-988-8381-59-3

For myself, I am interested in science and in philosophy only because I want to learn something about the riddle of the world in which we live, and the riddle of man's knowledge of that world. And I believe that only a revival of interest in these riddles can save the sciences and philosophy from narrow specialization and from an obscurantist faith in the expert's special skill, and in his personal knowledge and authority; a faith that so well fits our 'post-rationalist' and 'post-critical' age, proudly dedicated to the destruction of the tradition of rational philosophy, and of rational thought itself.

—*Karl Popper. The logic of scientific discovery. 1959.*

Short Contents

Table of Contents

1 Introduction

The GNU Astronomy Utilities (Gnuastro) is an official GNU package consisting of separate programs for the manipulation and analysis of astronomical data. See Appendix A [GNU Astronomy Utilities list], page 164 for the full list. All the various utilities share the same basic command line user interface for the comfort of both the users and developers. GNU Astronomy Utilities is written to comply fully with the GNU coding standards so it integrates finely with the GNU/Linux operating system. This also enables astronomers to expect a fully familiar experience in the source code, building, installing and command line user interaction that they have seen in all the other GNU software that they use.

For users who are new to the GNU/Linux environment, unless otherwise specified most of the topics in chapters 2 and 3 are common to all GNU software, for example installation, managing command line options or getting help. So if you are new to this environment, we encourage you to go through these chapters carefully. They can be a starting point from which you can continue to learn more from each program's own manual and fully enjoy this wonderful environment. This manual is written so someone who is completely new to GNU/Linux can get going very soon, see Section 1.6 [New to GNU/Linux?], page 5.

Finally it must be mentioned that in Gnuastro, no change to any program will be released before it has been fully documented in this manual first. As discussed in Section 1.2 [Science and its tools], page 2 this is the founding basis of the GNU Astronomy Utilities.

1.1 Quick start

Let's assume you have just downloaded the `gnuastro-0.1.tar.gz` in the `DOWLD` directory and you already have the requirements (see Section 3.1 [Requirements], page 21). Running the following commands will unpack, compile, check and install all the GNU Astronomy Utilities so you can use them anywhere in your system.

```
$ cd DOWLD
$ tar -zxvf gnuastro-0.1.tar.gz
$ cd gnuastro-0.1
$ ./configure
$ make
$ make check
$ sudo make install
$ cd ../
$ rm -rf gnuastro-0.1 gnuastro-0.1.tar.gz
```

See Section 3.3.4 [Known issues], page 29 if you confront any complications. For each program there is an 'Invoke ProgramName' sub-section in this manual which explains how the programs should be run on the command line. It can be read on the command line by running the command `$ info astprogname`, see Section 1.4 [Naming convention], page 4 and Section 4.6 [Getting help], page 43. The 'Invoke ProgramName' sub-section starts with a few examples of each program and goes on to explain the invocation details. In Chapter 2 [Tutorials], page 11 some real life examples of how these programs might be used is given.

1.2 Science and its tools

History of science indicates that there are always inevitably unseen faults, hidden assumptions, simplifications and approximations in all our theoretical models, data acquisition and analysis techniques. It is precisely these that will ultimately allow future generations to advance the existing experimental and theoretical knowledge through their new solutions and corrections.

In the past, scientists would gather data and process them individually to achieve an analysis thus having a much more intricate knowledge of the data and analysis. The theoretical models also required little (if any) simulations to compare with the data. Today both methods are becoming increasingly more dependent on pre-written software. Scientists are dissociating themselves from the intricacies of reducing raw observational data in experimentation or from bringing the theoretical models to life in simulations. These 'intricacies' are precisely those unseen faults, hidden assumptions, simplifications and approximations that define scientific progress.

> Unfortunately, most persons who have recourse to a computer for statistical analysis of data are not much interested either in computer programming or in statistical method, being primarily concerned with their own proper business. Hence the common use of library programs and various statistical packages. ...
> It's time that was changed.
> —F. J. Anscombe. The American Statistician, Vol. 27, No. 1. 1973

Anscombe's quartet[1] demonstrates how four data sets with widely different shapes (when plotted) give nearly identical output from standard regression techniques. Anscombe argues that "Good statistical analysis is not a purely routine matter, and generally calls for more than one pass through the computer". Anscombe's quartet can be generalized to say that users of a software cannot claim to understand how it works only based on the experience they have gained by frequently using it. This kind of subjective experience is prone to very serious mis-understandings about what it really does behind the scenes and can be misleading. This attitude is further encouraged through non-free software[2]. This approach to scientific software only helps in producing dogmas and an "obscurantist faith in the expert's special skill, and in his personal knowledge and authority"[3].

It is obviously impractical for any one human being to gain the intricate knowledge explained above for every step of an analysis. On the other hand, scientific data can be very large and numerous, for example images produced by telescopes in astronomy. This requires very efficient algorithms. To make things worse, natural scientists have generally not been trained in the advanced software techniques, paradigms and architecture that is taught in computer science or engineering courses and thus used in most software. The GNU Astronomy Utilities are an effort to tackle this issue. GNU Astronomy Utilities are built on the basis of the GNU general public license (GPL), giving the users complete "freedom" over them, see Section 1.3 [Your rights], page 3. We further add the requirement (on the authors of Gnuastro) that an astronomer, who is not necessarily trained in computer science or engineering, will need minimal requirements and preparations to understand and modify

---

[1] http://en.wikipedia.org/wiki/Anscombe%27s_quartet

[2] https://www.gnu.org/philosophy/free-sw.html

[3] Karl Popper. The logic of scientific discovery. 1959. Larger quote is given at the start of the PDF manual.

any step if they feel the need to do so, see Section 10.1 [Why C programming language?], page 148 and Section 10.2 [Design philosophy], page 150.

Imagine if Galileo did not have the technical knowledge to build a telescope. Astronomical objects could not be seen with the Dutch military design of the telescope. In the beginning of his "The Sidereal Messenger" (1610) he cautions the readers on this issue and instructs them on how to build a suitable instrument. Before he actually saw the moons of Jupiter, the mountains on the Moon or the crescent of Venus, he was an anti-Copernican and was "evasive" to Kepler[4]. Science is not independent of its tools.

Bjarne Stroustrup (creator of the C++ language) says: "Without understanding software, you are reduced to believing in magic". Ken Thomson (the designer or the Unix operating system) says "I abhor a system designed for the 'user' if that word is a coded pejorative meaning 'stupid and unsophisticated'." Certainly no scientist (user of a scientific software) would want to be considered as such. Roughly 5 years before special relativity and about two decades before quantum mechanics fundamentally changed Physics, Kelvin is quoted as saying[5]:

> There is nothing new to be discovered in physics now. All that remains is more and more precise measurement.
> —*William Thomson (Lord Kelvin), 1900*

If scientists are considered to be more than mere "puzzle solvers"[6], they cannot just passively sit back and wait for others to build the tools that form the basis of all their interpretations and working paradigms. Today there is a wealth of raw telescope images ready (mostly for free) at the finger tips of anyone who is interested with a fast enough internet connection to download them. The only thing lacking is new ways to analyze them and dig out the treasure that is lying hidden in them to existing methods and techniques.

> New data that we insist on analyzing in terms of old ideas (that is, old models which are not questioned) cannot lead us out of the old ideas. However many data we record and analyze, we may just keep repeating the same old errors, missing the same crucially important things that the experiment was competent to find.
> —*E. T. Jaynes, Probability theory, the logic of science. 2003.*

1.3 Your rights

The paragraphs below, in this section, belong to the GNU Texinfo[7] manual and are not written by us! The name "Texinfo" is just changed to "GNU Astronomy Utilities" or "Gnuastro" because they are released under the same licenses and it is beautifully written to inform you of your rights.

---

[4] Galileo G. (Translated by Maurice A. Finocchiaro). *The essential Galileo.* Hackett publishing company, first edition, 2008.

[5] Another such quote is from Albert. A. Michelson's speech at the dedication of Ryerson Physics Lab, U. of Chicago 1894 saying: "The more important fundamental laws and facts of physical science have all been discovered, and these are now so firmly established that the possibility of their ever being supplanted in consequence of new discoveries is exceedingly remote.... Our future discoveries must be looked for in the sixth place of decimals.".

[6] Thomas S. Kuhn. *The Structure of Scientific Revolutions*, University of Chicago Press, 1962.

[7] Texinfo is the GNU documentation system. It is used to create this manual in all the various formats.

GNU Astronomy Utilities is "free software"; this means that everyone is free to use it and free to redistribute it on certain conditions. Gnuastro is not in the public domain; it is copyrighted and there are restrictions on its distribution, but these restrictions are designed to permit everything that a good cooperating citizen would want to do. What is not allowed is to try to prevent others from further sharing any version of Gnuastro that they might get from you.

Specifically, we want to make sure that you have the right to give away copies of the programs that relate to Gnuastro, that you receive the source code or else can get it if you want it, that you can change these programs or use pieces of them in new free programs, and that you know you can do these things.

To make sure that everyone has such rights, we have to forbid you to deprive anyone else of these rights. For example, if you distribute copies of the Gnuastro related programs, you must give the recipients all the rights that you have. You must make sure that they, too, receive or can get the source code. And you must tell them their rights.

Also, for our own protection, we must make certain that everyone finds out that there is no warranty for the programs that relate to Gnuastro. If these programs are modified by someone else and passed on, we want their recipients to know that what they have is not what we distributed, so that any problems introduced by others will not reflect on our reputation.

The precise conditions of the licenses for the programs currently being distributed that relate to Gnuastro are found in the GNU General Public license that accompany them. This manual is covered by the GNU Free Documentation License.

1.4 Naming convention

GNU Astronomy Utilities is a package of independent utilities or programs. Each utility has an official name which consists of one or two words, describing what they do. The latter are printed with no space, for example NoiseChisel or ImageCrop. On the command line, you can run them with their executable names which start with an `ast` and might be an abbreviation of the official name, for example `astnoisechisel` or `astimgcrop`, see Section 3.3.1.3 [Executable names], page 27.

We will use "ProgramName" for a generic official program name and `astprogname` for a generic executable name. In this manual, the programs are classified based on what they do and thoroughly explained. An alphabetical list of the utilities that are installed on your system with this installation are given in Appendix A [GNU Astronomy Utilities list], page 164. That list also contains the executable names and version numbers along with a one line description.

1.5 Version numbering

The general Gnuastro package has a version number. It contains various programs and each of those also has its own version number. The version numbers for both are two numbers with a point (.) between them. The left number is the major version number while the right one is the minor version number. Note that the numbers are not decimals, so version 2.34 of a program is much more recent than version 2.5, which is not equal to 2.50!

The current version of Gnuastro is 0.1 and the version numbers of its various components are shown in Appendix A [GNU Astronomy Utilities list], page 164. To see the version of a program you are using, you can use the `--version` option, see Section 4.1.4 [Common options], page 34.

GNU Astronomy Utilities and all programs within it start with version number 0.1. For the programs, the minor version number is increased with any few bug fixes or small improvements which the developers decide is significant for a public release. So minor releases can be viewed as ad-hoc improvements. The major version number is set by a major goal which is defined by the developers of that particular program before hand.

For Gnuastro, its minor version number increases by 1 on every release (which contains an arbitrary number of updated version numbers for the programs or the general package). You can see the details from the `NEWS` file that comes with that distribution and is also available online to view before you download.

1.5.1 GNU Astronomy Utilities 1.0

The major version number of Gnuastro is increased similar to that of each program. Currently (prior to Gnuastro 1.0), the aim is to have a complete system for data manipulation and analysis at least similar to IRAF[8]. So an astronomer can take all the standard data analysis steps (starting from raw data to the final reduced product and standard post-reduction tools) with the various programs in Gnuastro.

The maintainers of each camera or detector on a telescope can provide a completely transparent shell script to the observer for data analysis. This script can set configuration files for all the required programs to work with that particular camera. The script can then run the proper programs in the proper sequence. The user/observer can easily follow the standard shell script to understand (and modify) each step and the parameters used easily. Bash (or other modern GNU/Linux shell scripts) are very powerful and made for this gluing job. This will simultaneously improve performance and transparency.

In order to achieve this and allow maximal creativity with the shell, the various programs have to be very low level programs and completely independent. Something like the GNU Coreutils.

1.6 New to GNU/Linux?

Some astronomers initially install and use the GNU/Linux operating systems because their research software can only be run in this environment. This is how the founder of Gnuastro started using GNU/Linux at least! If this is not the case for you, you can skip this section. Chapter 2 [Tutorials], page 11 is a complete chapter with some real world example applications of Gnuastro making good use of GNU/Linux capabilities written for newcomers to this environment. It is fully explained and is easy and entertaining to read, we hope you enjoy it.

You might have already noticed that we are not using the name "Linux", but "GNU/Linux". Please take the time to have a look at the following essays and FAQs for a complete understanding of this very important distinction.

- `https://www.gnu.org/gnu/gnu-users-never-heard-of-gnu.html`
- `https://www.gnu.org/gnu/linux-and-gnu.html`
- `https://www.gnu.org/gnu/why-gnu-linux.html`

---

[8] `http://iraf.noao.edu/`

- `https://www.gnu.org/gnu/gnu-linux-faq.html`

Another thing you will notice is that Gnuastro only has a command line user interface (CLI) or the 'shell' as it is referred to in Unix-like systems. This might be contrary to your mostly graphical user interface (GUI) experience with proprietary operating systems. To a first time user, the command line does appear much more complicated and adapting to it might not be easy.

Through GNOME 3[9], most GNU/Linux based operating systems now have a very advanced and useful GUI. Since the GUI was created long after the command line, some wrongly consider the command line to be obsolete. Both interfaces are very useful for different tasks (for example you can't view an image, video or web page on the command line!). Therefore they should not be regarded as rivals but as complementary, here we will outline how the CLI can be useful in scientific programs.

You can think of the GUI as a veneer over the CLI to facilitate a small subset of all the possible CLI operations. Each click you do on the GUI, can be thought of as internally running a different command. So asymptotically (if a good designer can design a GUI which is able to show you all the possibilities to click on) the GUI is only as powerful as the command line. In practice, such designers are very hard to come by for every program, so the GUI operations are always a subset of the internal CLI commands. For programs that are only made for the GUI, this results in not including lots of potentially useful operations. It also results in 'interface design' to be a crucially important part of any GUI program. Scientists don't usually have enough resources to hire a graphical designer, also the complexity of the GUI code is far more than CLI code, which is harmful for a scientific software, see Section 1.2 [Science and its tools], page 2.

For those operations with a GUI, one action on the GUI might be more efficient. However, if you have to repeat that same action more than once, it will soon become very frustrating and prone to errors. Unless the designers of a particular program decided to design such a system for a particular GUI action, there is no general way to run everything automatically on the GUI.

On the command line, with one command you can run numerous actions which can come from various CLI capable programs you have decided your self in any possible permutation with one command[10]. This allows for much more creativity than that offered to a GUI user. For technical and scientific operations, where the same operation (using various programs) has to be done on a large set of data files, this is crucially important. It also allows exact reproducability which is a foundation principle for scientific results. The most common CLI (which is also known as a shell) in GNU/Linux is GNU Bash, we strongly encourage you to put aside several hours and go through this beautifully explained web page: `https://flossmanuals.net/command-line/`. You don't need to read or even fully understand the whole thing, only a general knowledge of the first few chapters are enough to get you going.

Since the operations in the GUI are very limited and they are visible, reading a manual is not that important in the GUI (most programs don't even have any!). However, to give you the creative power explained above, with a CLI program, it is best if you first read the manual of any program you are using. You don't need to memorize any details, only an understanding of the generalities is needed. Once you start working, there are more easier

---

[9] `http://www.gnome.org/`

[10] By writing a shell script and running it, for example see the tutorials in Chapter 2 [Tutorials], page 11.

ways to remember a particular option or operation detail, see Section 4.6 [Getting help], page 43.

To experience the command line in its full glory and not in the GUI terminal emulator, press the following keys together: CTRL+ALT+F4[11] to access the virtual console. To return back to your GUI, press the same keys above replacing F4 with F1. In the virtual console, the GUI, with all its distracting colors and information, is gone. Enabling you to focus more accurately on your actual work.

For operations that use a lot of your system's resources (processing a large number of large astronomical images for example), the virtual console is the place to run them. This is because the GUI is not competing with your research work for your system's RAM and CPU. Since the virtual consoles are completely independent, you can even log out of your GUI environment to give even more of your hardware resources to the programs you are running and thus reduce the operating time.

Since it uses far less system resources, the CLI is also very convenient for remote access to your computer. Using secure shell (SSH) you can log in securely to your system (similar to the virtual console) from anywhere even if the connection speeds are low. There are apps for smart phones and tablets which allow you to do this.

1.7 Report a bug

According to Wikipedia "a software bug is an error, flaw, failure, or fault in a computer program or system that causes it to produce an incorrect or unexpected result, or to behave in unintended ways". So when you see that a program is crashing, not reading your input correctly, giving the wrong results, or not writing your output correctly, you have found a bug. In such cases, it is best if you report the bug to the developers. If it is an immediate issue, the developers will work hard to correct it as soon as possible.

Prior to actually filing a bug, it is best to search previous reports. The issue might have already been found and even solved. Recently corrected bugs are probably not yet publicly released because they are scheduled for the next Gnuastro stable release. If the bug is solved but not yet released and it is an urgent issue for you, you can get the version controlled source and compile that, see Section 10.4 [Version controlled source], page 152. There are generally two ways to inform us of bugs:

- Send a mail to bug-gnuastro@gnu.org. Any mail you send to this address will be distributed through the bug-gnuastro mailing list. This is the simplest and preferred way to send bug reports. The archives of this mailing list can be found at http://lists.gnu.org/archive/html/bug-gnuastro/.

- "Submit a new item" to the "Communication tools" section of the Gnuastro project webpage[12]. All the bug reports that are sent for Gnuastro (including the mailing list) are ultimately stored and managed in the Gnuastro project Bugs tracker[13]. Users can also initiate a bug report from the project webpage directly through the "Support"→"Submit new" links on the top of the page or the "Submit a new item"

---

[11] Instead of F4, you can use any of the keys from F2 to F6 for different virtual consoles. You can also run a separate GUI from within this console.

[12] https://savannah.gnu.org/projects/gnuastro/

[13] https://savannah.gnu.org/bugs/?group=gnuastro

link under the "Tech Support Manager" section (with a red question mark beside it) of the main Gnuastro project page. You can browse or submit a new item to this list anonymously. See Section 10.3 [Gnuastro project webpage], page 151 for more information about the central management hub of Gnuastro.

Once the items have been gathered from the mailing list or webpage, the developers will add it to either the "Bug Tracker" or "Task Manager" trackers of the Gnuastro project webpage. These two trackers can only be edited by the Gnuastro project members, but they can be browsed by anyone. So prior to filing a bug report please browse and search these two trackers to see if the issue has already been solved or is being solved.

> **Individual and independent bug reports:** If you have found multiple bugs, please send them as separate (and independent) mails as much as possible. This will significantly help us in managing and resolving them sooner.

> **Reproducible bug reports:** If we cannot reproduce your bug, then it is very hard to resolve it. So please send us a Minimal working example[14] along with the description. For example in running a program, please send us the full command line text and the output with the `-P` option, see Section 4.4 [Final parameter values, reproduce previous results], page 42. If it is caused only for a certain input, also send us that input file. In case the input FITS is large, please use ImageCrop to only crop the problematic section and make it as small as possible so it can easily be uploaded and downloaded and not waste the archive's storage, see Section 6.1 [ImageCrop], page 58.

1.8 Suggest new feature

We would always be very happy to hear of suggested new features. For every program there are already lists of features that we are planning to add. You can see the current list of plans from the Gnuastro project webpage at `https://savannah.gnu.org/projects/gnuastro/` and following "Tasks" → "Browse" at the top of the page. If you want to request a feature to an existing program, click on the "Display Criteria" above the list and under "Category", choose that particular program. Under "Category" you can also see the existing suggestions for new utilities or other cases like installation.

If the feature you want to suggest is not already listed in the task manager, then inform us through the `bug-gnuastro@gnu.org` mailing list or submitting an issue through the Gnuastro project webpage, see Section 1.7 [Report a bug], page 7. Please have in mind that the developers are all very busy with their own astronomical research, and implementing existing "task"s to add or resolving bugs. Gnuastro is a volunteer effort and none of the developers are paid for their hard work. So, although we will try our best, please don't not expect that your suggested feature be immediately included (with the next release of Gnuastro).

The best person to apply the feature is you, since you have the motivation and need. So you can read Chapter 10 [Developing], page 148 and start applying your desired feature. Once you have added it, you can use it for your own work and if you feel you want others

---

[14] `http://en.wikipedia.org/wiki/Minimal_Working_Example`

to benefit from your labour, you can request for it to become part of Gnuastro. You can then join the developers and start maintaining your own part (utility) of Gnuastro. If you choose to take this path of action please contact us before hand (Section 1.7 [Report a bug], page 7) so we can avoid possible duplicate activities and get interested people in contact.

> **Gnuastro is a collection of low level programs:** As described in Section 10.2 [Design philosophy], page 150, a founding principle of Gnuastro is that each program should be very basic and low-level. High level jobs should be done by running the separate programs in succession through a shell script, see the examples in Chapter 2 [Tutorials], page 11. So please consider how your desired job can best be broken into separate steps.

1.9 Announcements

Gnuastro has a dedicated mailing list for making announcements. Anyone that is interested can subscribe to this mailing list to stay upto date with new releases. To subscribe to this list, please visit `https://lists.gnu.org/mailman/listinfo/info-gnuastro`.

1.10 Conventions

In this manual we have the following conventions:

- All commands that are to be run on the shell (command line) prompt as the user start with a `$`. In case they must be run as a super-user or system administrator, they will start with a `#`. If the command is in a separate line and next line `is also in the code type face`, but doesn't have any of the `$` or `#` signs, then it is the output of the command after it is run. As a user, you don't need to type those lines.

- If the command becomes larger than the page width a `\` is inserted in the code. If you are typing the code by hand on the command line, you don't need to use multiple lines or add the extra space characters, so you can omit them. If you want to copy and paste these examples (highly discouraged!) then the `\` should stay.

  The `\` character is a shell escape character which is used commonly to make characters which have special meaning for the shell loose that special place (the shell will not treat them specially if there is a `\` behind them). When it is a last character in a line (the next character is a new-line charactor) the new-line character looses its meaning an the shell sees it as a simple white-space character, enabling you to use multiple lines to write your commands.

1.11 Acknowledgments

GNU Astronomy Utilities has significantly benefited from the help and support of various people and institutions. The plain text file `THANKS` which is distributed along with the source code has a full list. In particular the role of the Japanese Ministry of Science and Technology (MEXT) scholarship should be acknowledged for the long term scholarship of Mohammad Akhlaghi's Masters and PhD period in Tohoku University Astronomical Insitute in Sendai city. Gnuastro would not have been possible without the long term learning and planning that could only be acheived with such a long term scholarship. Tohoku University was the first institution to sign a copyright disclaimer to the Free Software Foundation for Gnuastro, allowing it to be freely available for the astronomical community. The very critical view points of Professor Takashi Ichikawa (at Tohoku University) were also instrumental in the creation of Gnuastro.

Mohammad-reza Khellat and Alan Lefor kindly studied the manual multiple times and provided very useful suggestions. Alan and Mohammad-reza also helped in testing Gnuastro on other operating systems. Brandon Invergo, Karl Berry and Richard Stallman also provided very useful suggestions during the GNU evaluation process. At first we wanted to submit the programs as independent and individual small programs, but thanks to their suggestions and ideas, all the separate programs were merged into the complete system that is now available for the astronomical community. Finally we should thank all the anonymous developers in various online forums which patiently answered all our small (but very important) technical questions.

2 Tutorials

In this chapter we give several tutorials or cookbooks on how to use the various tools in Gnuastro for your scientific purposes. In these tutorials, we have intentionally avoided too many cross references to make it more easily readable. To get more information about a particular program, you can visit the section with the same name as the program in this manual. Each program section starts by explaining the general concepts behind what it does. If you only want to see an explanation of the options and arguments of any program, see the subsection titled 'Invoking ProgramName'. See Section 1.10 [Conventions], page 9, for an explanation of the conventions we use in the example codes through the manual.

The tutorials in this section use a fictional setting of some historical figures in the history of astronomy. We have tried to show how Gnuastro would have been helpful for them in making their discoveries if there were GNU/Linux computers in their times! Please excuse us for any historical inaccuracy, this is not intended to be a historical reference. This form of presentation can make the tutorials more pleasent and entertaining to read while also being more practical (explaining from a user's point of view)[1]. The main reference for the historical facts mentioned in these fictional settings was Wikipedia.

2.1 Hubble visually checks and classifies his catalog

In 1924 Hubble[2] announced his discovery that some of the known nebulous objects are too distant to be within the the Milky Way (or Galaxy) and that they were probably distant Galaxies[3] in their own right. He had also used them to show that the redshift of the nebulae increases with their distance. So now he wants to study them more accurately to see what they actually are. Since they are nebulous or amorphous, they can't be modeled (like stars that are always a point) easily. So there is no better way to distinguish them than to visually inspect them and see if it is possible to classify these nebulae or not.

Hubble has stored all the FITS images of the objects he wants to visually inspect in his `/mnt/data/images` directory. He has also stored his catalog of extra Galactic nebulae in `/mnt/data/catalogs/extragalactic.txt`. Any normal user on his GNU/Linux system (including himself) only has read access to the contents of the `/mnt/data` directory. He has done this by running this command as root:

```
# chmod -R 755 /mnt/data
```

---

[1] This form of presenting a tutorial was influenced by the PGF/TikZ and Beamer manuals. The first provides graphic capabilities, while with the second you can make presentation slides in TeX and LaTeX. In these manuals, Till Tantau (author of the manual) uses Euclid as the protagonist. There are also some nice words of wisdom for Unix-like systems called "Rootless Root": http://catb.org/esr/writings/unix-koans/. These also have a similar style but they use a mythical figure named Master Foo. If you already have some experience in Unix-like systems, you will definitely find these "Unix Koans" very entertaining.

[2] Edwin Powell Hubble (1889 – 1953 A.D.) was an American astronomer who can be considered as the father of extragalactic astronomy, by proving that some nebulae are too distant to be within the Galaxy. He then went on to show that the universe appears to expand and also done a visual classification of the galaxies that is known as the Hubble fork.

[3] Note that at that time, "Galaxy" was a proper noun used to refer to the Milky way. The concept of a galaxy as we define it today had not yet become common. Hubble played a major role in creating today's concept of a galaxy.

Hubble has done this intentionally to avoid mistakenly deleting or modifying the valuable images he has taken at Mount Wilson while he is working as an ordinary user. Retaking all those images and data is simply not an option. In fact they are also in another hard disk (/dev/sdb1). So if the hard disk which stores his GNU/Linux distribution suddenly malfunctions due to work load, his data is not in harms way. That hard disk is only mounted to this directory when he wants to use it with the command:

```
# mount /dev/sdb1 /mnt/data
```

In short, Hubble wants to keep his data safe and fortunately by default Gnuastro allows for this. Hubble creates a temporary visualcheck directory in his home directory for this check. He runs the following commands to make the directory and change to it[4]:

```
$ mkdir ~/visualcheck
$ cd ~/visualcheck
$ pwd
/home/edwin/visualcheck
$ ls
```

Hubble has multiple images in /mnt/data/images, some of his targets might be on the edges of an image and so several images need to be stitched to give a good view of them. Also his extra Galactic targets belong to various pointings in the sky, so they are not in one large image. Gnuastro's ImageCrop is just the utility he wants. The catalog in extragalactic.txt is a plain text file which stores the basic information of all his known 200 extra Galactic nebulae. In its second column it has each object's Right Ascension (the first column is a label he has given to each object) and in the third the object's declination. Having read the Gnuastro manual, he knows that all counting is done starting from zero, so the RA and Dec columns have number 1 and 2 respectively.

```
$ astimgcrop --racol=1 --deccol=2 /mnt/data/images/*.fits      \
             /mnt/data/catalogs/extragalactic.txt
ImageCrop started on Tue Jun  14 10:18:11 1932
   ---- ./4_crop.fits                 1 1
   ---- ./2_crop.fits                 1 1
   ---- ./1_crop.fits                 1 1
[[[ Truncated middle of list ]]]
   ---- ./198_crop.fits               1 1
   ---- ./195_crop.fits               1 1
  - 200 images created.
  - 200 were filled in the center.
  - 0 used more than one input.
ImageCrop finished in:  2.429401 (seconds)
```

Hubble already knows that thread allocation to the the CPU cores is asynchronous, so each time you run it the order of which job gets done first differs. When using ImageCrop the order of outputs is irrelevant since each crop is independent of the rest. This is why the crops are not necessarily created in the same input order. He is content with the default width of the outputs (which he inspected by running $ astimgcrop -P). If he wanted a different width for the cropped images, he could do that with the --wwidth option which

---

[4] The pwd command is short for "Print Working Directory" and ls is short for "list" which shows the contents of a directory.

accepts a value in arcseconds. When he lists the contents of the directory again he finds his 200 objects as separate FITS images.

```
$ ls
1_crop.fits 2_crop.fits ... 200_crop.fits
```

The FITS image format was not designed for viewing, but mainly for accurate storing of the data. So he chooses to convert the cropped images to a more common image format to view them more quickly and easily through standard image viewers (which load much faster than FITS image viewer). JPEG is one of the most recognized image formats that is supported by most image viewers. Fortuantely Gnuastro has just such a tool to convert various types of file types to and from each other: ConvertType. Hubble has already heard of GNU Parallel from one of his colleagues at Mount Wilson Observatory. It allows multiple instances of a command to be run simultaneously on the system, so he uses it in conjuction with ConvertType to convert all the images to JPEG.

```
$ parallel astconvertt -ojpg ::: *_crop.fits
```

For his graphical user interface Hubble is using GNOME which is the default in most distributions in GNU/Linux. The basic image viewer in GNOME is the Eye of GNOME, which has the executable file name eog[5]. Since he has used it before, he knows that once it opens an image, he can use the **ENTER** or **SPACE** keys on the keyboard to go to the next image in the directory or the **Backspace** key to to go the previous image. So he opens the image of the first object with the command below and with his cup of coffee in his other hand, he flips through his targets very fast to get a good initial impression of the morphologies of these extra Galactic nebulae.

```
$ eog 1_crop.jpg
```

Hubble's cup of coffee is now finished and he also got a nice general impression of the shapes of the nebulae. He tentatively/mentally classified the objects into three classes while doing the visual inspection. One group of the nebulae have a very simple elliptical shape and seem to have no internal special structure, so he gives them code 1. Another clearly different class are those which have spiral arms which he associates with code 2 and finally there seems to be a class of nebulae in between which appear to have a disk but no spiral arms, he gives them code 3.

Now he wants to know how many of the nebulae in his extra Galactic sample are within each class. Repeating the same process above and writing the results on paper is very time consuming and prone to errors. Fortunately Hubble knows the basics of GNU Bash shell programming, so he writes the following short script with a loop to help him with the job. After all, computers are made for us to operate and knowing basic shell programming gives Hubble this ability to creatively operate the computer as he wants. So using GNU Emacs[6] (his favorite text editor) he puts the following text in a file named **classify.sh**.

```
for name in *.jpg
do
    eog $name &
    processid=$!
```

---

[5] Eye of GNOME is only available for users of the GNOME graphical desktop environment which is the default in most GNU/Linux distributions. If you use another graphical desktop environment, replace **eog** with any other image viewer.

[6] This can be done with any text editor

```
        echo -n "$name belongs to class: "
        read class
        echo $name $class >> classified.txt
        kill $processid
    done
```

Fortunately GNU Emacs or even simpler editors like Gedit (part of the GNOME graphical user interface) will display the variables and shell constructs in different colors which can really help in understanding the script. Put simply, the `for` loop gets the name of each JPEG file in the directory this script is run in and puts it in `name`. In the shell, the value of a variable is used by putting a `$` sign before the variable name. Then Eye of GNOME is run on the image in the background to show him that image and its process ID is saved internally (this is necessary to close Eye of GNOME later). The shell then prompts the user to specify a class and after saving it in `class`, it prints the file name and the given class in the next line of a file named `classified.txt`. To make the script executable (so he can run it later any time he wants) he runs:

```
$ chmod +x classify.sh
```

Now he is ready to do the classification, so he runs the script:

```
$ ./classify.sh
```

In the end he can delete all the JPEG and FITS files along with ImageCrop's log file with the following short command. The only files remaining are the script and the result of the classification.

```
$ rm *.jpg *.fits astimgcrop.txt
$ ls
classified.txt    classify.sh
```

He can now use `classified.txt` as input to a plotting program to plot the histogram of the classes and start making interpretations about what these nebulous objects that are outside of the Galaxy are.

2.2 Sufi simulates a detection

It is the year 953 A.D. and Sufi[7] is in Shiraz as a guest astronomer. He had come there to use the advanced 123 centimeter astrolabe for his studies on the Ecliptic. However, something was bothering him for a long time. While mapping the constellations, there were several non-stellar objects that he had detected in the sky, one of them was in the Andromeda constellation. During a trip he had to Yemen, Sufi had seen another such object in the southern skies looking over the indian ocean. He wasn't sure if such cloud-like non-stellar objects (which he was the first to call 'Sahābi' in Arabic or 'nebulous') were real astronomical objects or if they were only the result of some bias in his observations. Could such diffuse objects actually be detected at all with his detection technqiue?

He still had a few hours left until nightfall (when he would continue his studies on the ecliptic) so he decided to find an answer to this question. He had thoroughly studied Claudius Ptolemy's (90 – 168 A.D) Almagest and had made lots of corrections to it, in

---

[7] Abd al-rahman Sufi (903 – 986 A.D.), also known in Latin as Azophi was an Iranian astronomer. His manuscript "Book of fixed stars" contains the first recorded observations of the Andromeda galaxy, the Large Magellanic Cloud and seven other non-stellar or 'nebulous' objects.

particular in measuring the brightness. Using his same experience, he was able to measure a magnitude for the objects and wanted to simulate his observation to see if a simulated object with the same brightness and size could be detected in a simulated noise with the same detection technique. The general outline of the steps he wants to take are:

1. Make some mock profiles in an oversampled image. The initial mock image has to be oversampled prior to convolution or other forms of transformation in the image. Through his experiences, Sufi knew that this is because the image of heavenly bodies is actually transformed by the atmosphere or other sources outside the atmosphere (for example gravitational lenses) prior to being sampled on an image. Since that transformation occurs on a continuous grid, to best approximate it, he should do all the work on a finer pixel grid. In the end he can resample the result to the initially desired grid size.

2. Convolve the image with a PSF image that is oversampled to the same value as the mock image. Since he wants to finish in a reasonable time and the PSF kernel will be very large due to oversampling, he has to use frequency domain convolution which has the side effect of dimming the edges of the image. So in the first step above he also has to build the image to be larger by at least half the width of the PSF convolution kernel on each edge.

3. With all the transformations complete, the image should be resampled to the same size of the pixels in his detector.

4. He should remove those extra pixels on all edges to remove frequency domain convolution artifacts in the final product.

5. He should add noise to the (until now, noise-less) mock image. After all, all observations have noise associated with them.

Fortunately Sufi had heard of GNU Astronomy Utilities from a colleague in Isfahan (where he worked) and had installed it on his computer a year before. It had tools to do all the steps above. He had used MakeProfiles before, but wasn't sure which columns he had chosen in his user or system wide configuration files for which parameters, see Section 4.2 [Configuration files], page 37. So to start his simulation, Sufi runs MakeProfiles with the -P option to make sure what columns in a catalog MakeProfiles currently recognizes and the output image parameters:

```
$ astmkprof -P
# MakeProfiles (GNU Astronomy Utilities 0.1) 0.1
# Configured on 21 September 952 at 19:37
# Written on Sat Oct  6 15:49:31 953

# Output:
 naxis1             1000
 naxis2             1000
 oversample         5

[[[ Truncated middle of list ]]]

# Catalog:
 xcol               1
```

```
ycol            2
fcol            3
rcol            4
ncol            5
pcol            6
qcol            7
mcol            8
tcol            9
```

```
[[[ Truncated rest of list ]]]
```

In particular, Sufi looks at the parameters under the catalog grouping. Fortunately the columns are naturally numbered such that column 0 can be an ID he specifies for each object (which MakeProfiles ignores) and each subsequent column specifies a given parameter. Fortunately MakeProfiles has the capability to also make the PSF which is to be used on the mock image and using the --prepforconv option, he can also make the mock image to be larger by the correct amount and all the sources to be shifted by the correct amount.

For his initial check he decides to simulate the nebula in the Andromeda constellation. The night he was observing, the PSF had roughly a FWHM of about 5 pixels, so as the first row, he defines the PSF parameters and sets the radius column (rcol above, fifth column) to 5.000, he also chooses a Moffat function for its functional form. Remembering how diffuse the nebula in the Andromeda constellation was, he decides to simulate it with a mock Sérsic index 1.0 profile. He wants the output to be 500 pixels by 500 pixels, so he puts the mock profile in the center. Looking at his drawings of it, he decides a reasonable effective radius for it would be 40 pixels on this image pixel scale, he sets the axis ratio and position angle to approximately correct values too and finally he sets the total magnitude of the profile to 3.44 which he had accurately measured. Sufi also decides to truncate both the mock profile and PSF at 5 times the respective radius parameters. In the end he decides to put four stars on the four corners of the image at very low magnitudes as a visual scale.

Using all the information above, he creates the catalog of mock profiles he wants in a file named cat.txt (short for catalog) using his favorite text editor and stores it in a directory named simulationtest in his home directory[8]:

```
$ mkdir ~/simulationtest
$ cd ~/simulationtest
$ pwd
/home/rahman/simulationtest
$ emacs cat.txt
$ ls
cat.txt
$ cat cat.txt
 0   0.0000   0.0000   1   5.000   4.765   0.0000   1.000   30.000   5.000
 1   250.00   250.00   0   40.00   1.000  -25.00   0.400   3.4400   5.000
 2   50.000   50.000   3   0.000   0.000   0.0000   0.000   9.0000   0.000
 3   450.00   50.000   3   0.000   0.000   0.0000   0.000   9.2500   0.000
 4   50.000   450.00   3   0.000   0.000   0.0000   0.000   9.5000   0.000
```

---

[8] The cat command prints the contents of a file, short for concatenation.

```
5  450.00   450.00  3  0.000  0.000  0.0000  0.000  9.7500  0.000
```

He looked into his observation logs and found that the night he was observing, the zeropoint magnitude was 18. Now he has all the necessary parameters and runs MakeProfiles with the following command:

```
$ astmkprof --prepforconv --naxis1=500 --naxis2=500            \
            --zeropoint=18.0 cat.txt
MakeProfiles started on Sat Oct  6 16:26:56 953
  - 6 profiles read from cat.txt          in 0.000209 seconds
  ---- Row 5 complete, 5 left to go.
  ---- Row 3 complete, 4 left to go.
  ---- Row 2 complete, 3 left to go.
  ---- Row 4 complete, 2 left to go.
  ---- ./0.fits created.
  ---- Row 0 complete, 1 left to go.
  ---- Row 1 complete, 0 left to go.
  - cat.fits created.                     in 0.024811 seconds
MakeProfiles finished in:  0.236629 (seconds)
$ls
0.fits  astmkprof.log  cat.fits  cat.txt
```

The file 0.fits is the PSF Sufi had asked for and cat.fits is the image containing the 5 objects. The PSF is now available to him as a separate file for the convolution step. While he was preparing the catalog, one of his students came up and was also following the steps. When he opened the image, the student was surprised to see that all the stars are only one pixel and not in the shape of the PSF as we see when we image the sky at night. So Sufi explained to him that the stars will take the shape of the PSF after convolution and this is how they would look if we didn't have an atmosphere or an aperture when we took the image. The size of the image was also surprising for the student, instead of 500 by 500, it was 2630 by 2630 pixels. So Sufi had to explain why oversampling is very important for parts of the image where the flux change is significant over a pixel. Sufi then explained to him that after convolving we will resample the image to get our originally desired size. To convolve the image, Sufi ran the following command:

```
$ astconvolve --kernel=0.fits cat.fits
Convolve started on Mon Apr  6 16:35:32 953
Convolving cat.fits (hdu: 0)
 with the kernel 0.fits (hdu: 0).
 using 8 CPU threads in the frequency domain.
  - Input and Kernel images padded.        in 0.045576 seconds
  - Images converted to frequency domain.  in 10.486712 seconds
  - Multiplied in the frequency domain.    in 0.032780 seconds
  - Converted back to the spatial domain.  in 5.342335 seconds
  - Padded parts removed.                  in 0.011880 seconds
Convolve finished in:  15.972771 (seconds)
$ls
0.fits  astmkprof.log  cat_convolved.fits  cat.fits  cat.txt
```

When convolution finished, Sufi opened the `cat_convolved.fits` file and showed the effect of convolution to his student and explained to him how a PSF with a larger FWHM would make the points even wider. With the convolved image ready, they were ready to re-sample it to the orignal pixel scale Sufi had planned. Sufi explained the basic concepts of warping the image to his student and also the fact that since the center of a pixel is assumed to take integer values in the FITS standard, the transformation matrix would not be a simple scaling but would also need translating, see Section 6.3.2 [Merging multiple warpings], page 89. Then he ran ImageWarp with the following command:

```
$ astimgwarp cat_convolved.fits --matrix="0.2,0,0.4  0,0.2,0.4  0,0,1"
ImageWarp started on Mon Apr  6 16:51:59 953
ImageWarp finished in:  0.481421 (seconds)
$ ls
0.fits           cat_convolved.fits           cat.fits
astmkprof.log  cat_convolved_warped.fits  cat.txt
```

`cat_convolved_warped.fits` now has the correct pixel scale. However, the image is still larger than what we had wanted, it is 526 (500 + 13 + 13) by 526 pixels. The student is slightly confused, so Sufi also resamples the PSF with ImageWarp and the same warping matrix and shows him that it is 27 ($2 \times 13 + 1$) by 27 pixels. Sufi goes on to explain how frequency space convolution will dim the edges and that is why he added the `--prepforconv` option to MakeProfiles, see Section 8.1.2 [If convolving afterwards], page 134. Now that convolution is done Sufi can remove those extra pixels using ImageCrop:

```
$ astimgcrop cat_convolved_warped.fits --section=13:*-13,13:*-13
ImageCrop started on Sat Oct  6 17:03:24 953
  - Read metadata of 1 images.              in 0.000560 seconds
  ---- cat_convolved_warped_crop.fits 1 1
ImageCrop finished in:  0.018917 (seconds)
$ls
0.fits           astmkprof.log           cat_convolved_warped.fits
0_warped.fits  cat_convolved.fits           cat.fits
astimgcrop.log  cat_convolved_warped_crop.fits  cat.txt
```

Finally, the `cat_convolved_warped.fits` has the same dimensionality as Sufi had asked for in the beginning. All this trouble was certainly worth it because now there is no dimming on the edges of the image and the profile centers are more accurately sampled. The final step to simulate a real observation would be to add noise to the image. Sufi set the zeropoint magnitude to the same value that he set when making the mock profiles and looking again at his observation log, he found that at that night the background flux near the nebula had a magnitude of 7. So using these values he ran MakeNoise:

```
$ astmknoise --zeropoint=18 --background=7 --output=out.fits     \
            cat_convolved_warped_crop.fits
MakeNoise started on Mon Apr  6 17:05:06 953
  - Generator type: mt19937
  - Generator seed: 1428318100
MakeNoise finished in:  0.033491 (seconds)
$ls
0.fits           cat_convolved.fits                cat.txt
0_warped.fits  cat_convolved_warped_crop.fits  out.fits
```

```
astimgcrop.log   cat_convolved_warped.fits
astmkprof.log    cat.fits
```

The `out.fits` file now has the noised image of the mock catalog Sufi had asked for. Seeing how the `--output` option allows the user to specify the name of the output file, the student was confused and wanted to know why Sufi hadn't used it before? Sufi then explained to him that for intermediate steps it is best to rely on the automatic output, see Section 4.5 [Automatic output], page 43. Doing so will give all the intermediate files the same basic name structure, so in the end you can simply remove them all with the Shell's capabilities. So Sufi decided to show this to the student by making a shell script from the commands he had used before.

The command line shell has the capability to read all the separate input commands from a file. This is very useful when you want to do the same thing multiple times, with only the names of the files or minor parameters changing between the different instances. Using the shell's history (by pressing the up keyboard key) Sufi reviewed all the commands and then he retrieved the last 5 commands with the `$ history 5` command. He selected all those lines he had input and put them in a text file named `mymock.sh`. Then he used some shell variables to set the two main constant parts of all the command to generalized variables.

```
edge=13
base=cat
rm out.fits

astmkprof --prepforconv --naxis1=500 --naxis2=500            \
          --zeropoint=18.0 "$base".txt
astconvolve --kernel=0.fits "$base".fits
astimgwarp "$base"_convolved.fits --matrix="0.2,0,0.4  0,0.2,0.4  0,0,1"
astimgcrop "$base"_convolved_warped.fits                     \
          --section=$edge:*-$edge,$edge:*-$edge
astmknoise --zeropoint=18 --background=7 --output=out.fits    \
          "$base"_convolved_warped_crop.fits
rm 0*.fits cat*.fits *.log
```

Sufi then explained to the eager student that you define a variable by giving it a name, followed by an = sign and the value you want. Then you can reference that variable from anywhere in the script by calling its name with a `$` prefix. So in the script whenever you see `$base`, the value we defined for it above is used. If you use advanced editors like GNU Emacs or even simpler ones like Gedit (part of the GNOME graphical user interface) the variables will become a different color which can really help in understanding the script. We have put all the `$base` variables in double quotation marks (`"`) so the variable name and the following text do not get mixed, the shell is going to ignore the `"` after replacing the variable value. To make the script executable, Sufi ran the following command:

```
$ chmod +x mymock.sh
```

Then finally, Sufi ran the script, simply by calling its file name:

```
$ ./mymock.sh
```

After the script finished, the only file remaining is the `out.fits` file that Sufi had wanted in the beginning. Sufi then explained to the student how he could run this script anywhere that he has a catalog if the script is in the same directory. The only thing the student had to modify in the script was the name of the catalog (the value of the `base` variable in the start of the script) and the value to the `edge` variable if he changed the PSF size. The student was also very happy to hear that he won't need to make it executable again when he makes changes later, it will remain executable unless he explicitly changes the executable flag with `chmod`.

The student was really excited, since now, through simple shell scripting, he could really speed up his work and run any command in any fashion he likes allowing him to be much more creative in his works. Until now he was using the graphical user interface which doesn't have such a facility and doing repetitive things on it was really frustrating and some times he would make mistakes. So he left to go and try scripting on his own computer.

Sufi could now get back to his own work and see if the simulated nebula which resembled the one in the Andromeda constellation could be detected or not. Although it was extremely faint[9], fortunately it passed his detection tests and he wrote it in the draft manuscript that would later become "Book of fixed stars". He still had to check the other nebula he saw from Yemen and several other such objects, but they could wait until tomorrow (thanks to the shell script, he only has to define a new catalog). It was nearly sunset and they had to begin preparing for the night's measurements on the ecliptic.

---

[9] The brightness of a diffuse object is added over all its pixels to give its final magnitude, see Section 8.1.3 [Flux Brightness and magnitude], page 134. So although the magnitude 3.44 (of the mock nebula) is orders of magnitude brighter than 6 (of the stars), the central galaxy is much fainter. Put another way, the brightness is distributed over a large area in the case of a nebula.

3 Installation

To successfully install Gnuastro you have to have the requirements already installed on your system. They are very basic for most astronomical programs and you might already have them installed. To check, try running the `$ ./configure` script. If you get no errors, then you already have them and you can skip Section 3.1 [Requirements], page 21. You can heavily customize your install of Gnuastro, to learn more about them, see Section 3.3 [Installing GNU Astronomy Utilities], page 23. If you encounter any problems in the installation process, it is probably already explained in Section 3.3.4 [Known issues], page 29. In Appendix B [Other useful software], page 165 the installation and usage of some other free software that are not directly required by Gnuastro but might be useful in conjunction with it is discussed.

3.1 Requirements

GNU Astronomy Utilities 0.1 have several dependencies, they all follow the same basic GNU based build system (like that shown in Section 1.1 [Quick start], page 1), so even if you don't have them, installing them should be pretty straightforward. In this section we explain each program and any specific note that might be necessary in the installation.

The most basic choice is to build the packages from source your self instead of relying on your distribution's pre-built packages. These packages might already be available by your distribution's package management system. You can also use those, just note the following two issues:

1. They might not be the most recent release.

2. For each package, Gnuastro might require certain configuration options that the your distribution's package managers didn't add for you. Those configuration options are explained below.

3. For the libraries, they might separate the binary file from the header files, see Section 3.3.4 [Known issues], page 29.

3.1.1 GNU Scientific library

The GNU Scientific Library is probably already present in your distribution's package management system. To install it from source, you can run the following commands after you have downloaded[1] `gsl-X.X.tar.gz`:

```
$ tar -zxvf gsl-X.X.tar.gz
$ cd gsl-X.X
$ ./configure
$ make
$ make check
$ sudo make install
```

---

[1] http://www.gnu.org/software/gsl/

3.1.2 CFITSIO

CFITSIO is the closest you can get to the pixels in a FITS image while remaining faithful to the FITS standard[2]. It is written by William Pence, the author of the FITS standard[3], and is regularly updated. Setting the definitions for all other software packages using FITS images.

Some GNU/Linux distributions have CFITSIO in their package managers, if it is available and updated, you can use it. One problem that might occur is that CFITSIO might not be configured with the `--enable-reentrant` option by the distribution. This option allows CFITSIO to open a file in multiple threads. If so, upon running, any program which needs this capability will warn you and abort if you ask for multiple threads. In such cases you can take the following step.

The best way is that you can install CFITSIO from source. You can download the latest version of the source code and manual from its webpage[4]. We strongly recommend that you have a look through Chapter 2 (Creating the CFITSIO library) of the CFITSIO manual and understand the options you can pass to `$ ./configure` (they aren't too much). This is a very basic package for most astronomical software and it is best that you configure it nicely with your system. Once you download the source and unpack it, the following configure script should be enough for most purposes. Don't forget to read chapter two of the manual though, for example the second option is only for 64bit systems. The manual also explains how to check if it has been installed correctly.

```
$ tar -vxzf cfitsio_latest.tar.gz
$ cd cfitsio
$ ./configure --prefix=/usr/local --enable-sse2 --enable-reentrant
$ make
$ sudo make install
```

3.1.3 WCSLIB

WCSLIB is also written and maintained by one of the authors of the World Coordinate System (WCS) definition in the FITS standard[5], Mark Calabretta. It might be already built and ready in your distribution's package management system. Here installation from source is explained. To install WCSLIB you will need to have CFITSIO already installed, see Section 3.1.2 [CFITSIO], page 22. WCSLIB also has plotting capabilities which use PGPLOT (a plotting library for C). However, if you will not be using its plotting functions, you can configure it such that pgplot is not required.

If you do want to make plots with WCSLIB, there is an explanation in Section B.2 [PGPLOT], page 166. To disable the dependency on PGPLOT, you have to add the --without-pgplot option to the configure script as you can see below. You can get the most recent source code from the WCSLIB webpage[6]. In the directory where you have

---

[2] http://fits.gsfc.nasa.gov/fits_standard.html

[3] Pence, W.D. et al. Definition of the Flexible Image Transport System (FITS), version 3.0. (2010) Astronomy and Astrophysics, Volume 524, id.A42, 40 pp.

[4] http://heasarc.gsfc.nasa.gov/fitsio/fitsio.html

[5] Greisen E.W., Calabretta M.R. (2002) Representation of world coordinates in FITS. Astronomy and Astrophysics, 395, 1061-1075.

[6] http://www.atnf.csiro.au/people/mcalabre/WCS/

downloaded the compressed file, you can take the following steps (the `x.xx` represents the version number):

```
$ tar -jxvf wcslib.tar.bz2
$ cd wcslib-x.xx
$ ./configure --without-pgplot LIBS="-pthread -lm"
$ make
$ make check
$ sudo make install
```

3.2 Optional requirements

Most of the programs in Gnuastro make use of the libraries in Section 3.1 [Requirements], page 21, therefore if they are not available, the configure script will complain and compiling the Gnuastro is not possible. The libraries listed in this section are only used for very specific applications, therefore if you don't want these operations, they do not need to be present.

If the `./configure` script can't find these requirements, it will warn you that they are not present and notify you of the operation(s) you can't do due to not having them. If the output you request from a program requires a missing library, that program is going to warn you and abort. In the case of executables like GPL GhostScript, if you install them at a later time, the program will run. This is because if required libraries are not present at build time, the executables cannot be built, but an executable is called by the built program at run time so if it becomes available, it will be used. If you do install an optional library later, you will have to rebuild Gnuastro and reinstall it for it to take effect.

3.2.1 libjpeg

libjpeg is only used by ConvertType to read from and write to JPEG images. libjpeg is a very basic library that provides tools to read and write JPEG images, most of the GNU/Linux graphic programs and libraries use it. Therefore you most probably already have it installed. libjpeg-turbo is an alternative to libjpeg. It uses SIMD instructions for ARM based systems that significantly decreases the processing time of JPEG compression and decompression algorithms.

3.2.2 GPL Ghostscript

GPL Ghostscript's executable (`gs`) is called used by ConvertType to compile a PDF file from a source PostScript file, see Section 5.2 [ConvertType], page 51. Therefore its headers (and libraries) are not needed. With a very high probability you already have it in your GNU/Linux distribution. Unfortunately it does not follow the standard GNU build style so installing it is very hard. It is best to rely on your distribution's package managers for this.

3.3 Installing GNU Astronomy Utilities

This section is basically a longer explanation to the sequence of commands given in Section 1.1 [Quick start], page 1. If you want to have all the programs of Gnuastro installed in your system, you don't want to change the executable names during or after installation, you have root access to install the programs in a system wide directory, the

Letter paper size of the print manual is fine for you or as a summary you don't feel like going into the details when everything is working seamlessly, you can safely skip this section. If you have any of the above problems or you want to understand the details for a better control over your build and install, read along.

In the following it is assumed that you have downloaded the compressed source file, `gnuastro-0.1.tar.gz`, to the `DOWLD` (short for download) directory, replace this name with the directory that you want to run the installation in. Note that after installation, if you don't plan to re-install you no longer need this file or the uncompressed directory, so you can safely delete both. The first three steps in Section 1.1 [Quick start], page 1 need no extra explanation. Once you uncompress the source file the directory `DOWLD/gnuastro-0.1` will be created.

3.3.1 Configuring

The `$ ./configure` step is the most important step in the build and install process. All the required packages, libraries, headers and environment variables are checked in this step. The behaviors of make and make install can also be set through command line options to this command.

The configure script accepts various arguments and options which enable the final user to highly customize whatever she is building. The options to configure are generally very similar to normal program options explained in Section 4.1.1 [Arguments and options], page 31. Similar to all GNU programs, you can get a full list of the options along with a short explanation by running

        `$ ./configure --help`

A complete explanation is also included in the `gnuastro-0.1/INSTALL` file in plain text that comes with the Gnuastro source. Note that this file was written by the authors of GNU Autoconf and is common for all programs which use the `$ ./configure` script for building and installing (there is a lot of such programs). Here the most common general usages (not only Gnuastro) are explained: when you don't have super-user access to the system and changing the executable names. But before that a review of the options to configure that are particular to Gnuastro are discussed.

3.3.1.1 GNU Astronomy Utilities configure options

Most of the options to configure (which are to do with building) are similar for every program which uses this script. Here the options that are particular to Gnuastro are discussed. The next topics explain the usage of other configure options which can be applied to any program using the GNU build system (through the configure script).

`--with-numthreads`

> (=`INT`) If this option is given an integer value, that value will be used for the default number of threads to use. If it is not given, then the total number of threads will be read from the system, see Section 4.3 [Threads in GNU Astronomy Utilities], page 40. Specifying `--with-numthreads=no` or `--without-numthreads` is equivalent to not calling this option it at all.

`--enable-progname`

> Only build and install `progname` along with any other program that is enabled in this fashion. `progname` is the name of the executable without the `ast`, for

example `imgcrop` for ImageCrop (with the executable name of `astimgcrop`). If this option is called for any of the programs in Gnuastro, any program which is not explicitly enabled will not be built or installed.

`--disable-progname`
`--enable-progname=no`

> Do not build or install the program named `progname`. This is very similar to the `--enable-progname`, but will build and install all the other programs except this one.

`--enable-gnulibcheck`

> Enable checks on the GNU Portability Library (Gnulib). Gnulib is used by Gnuastro to enable users of non-GNU based operating systems (that don't use GNU C Library or glibc) to compile and use the advanced features that this library provides. We make extensive use of such functions. If you give this option to `$ ./configure`, when you run `$ make check`, first the functions in Gnulib will be tested, then the Gnuastro executables. If your operating system does not support glibc or has an older version of it and you have problems in the build process (`$ make`), you can give this flag to configure to see if the problem is caused by Gnulib not supporting your operating system or Gnuastro, see Section 3.3.4 [Known issues], page 29.

> **Note:** If some programs are enabled and some are disabled, it is equivalent to simply enabling those that were enabled. Listing the disabled programs is redundant.

Note that the tests of some programs might require other programs to have been installed and tested. For example MakeProfiles is the first program to be tested when you run `$ make check`, it provides the inputs to all the other tests. So if you don't install MakeProfiles, then the tests for all the other programs will be skipped or fail. To avoid this, in one run, you can install all the packages and run the tests but not install. If everything is working correctly, you can run configure again with only the packages you want but not run the tests and directly install after building.

3.3.1.2 Installation directory

One of the most commonly used options to configure is the directory that will host all the files which require installing, for example the actual executable files for the program, the documentation and configuration files. This is done through the `--prefix` option. To demonstrate its applicability, let's assume you don't have root access to the computer you are using which is one of the most common usage cases.

In case you don't have super user or root access to the system, you can't take the installation steps of the command sequence in Section 1.1 [Quick start], page 1. To be able to access the Gnuastro executable files from anywhere, you have to specify a special directory in the directories you have write access in, through the shell's environment variables. Note that this explanation can apply to all the requirements in Section 3.1 [Requirements], page 21 in case the system lacks them or the system wide install was not built with the proper configuration options. We will start with a short introduction to the shell variables.

Shell variable values are basically treated as strings of characters. You can define a variable and a value for it by running `$ myvariable=a test value` on the command line. Then you can see the value in the with the command `$ echo $myvariable`. If a variable has no value, this command will only print an empty line. This variable will be known as long as this shell or terminal is running. Other terminals will have no idea it existed. The main advantage of shell variables is that if they are exported[7] subsequent programs in that shell can access their value. So by setting them to any desired value, you can change the 'environment' of the program. The shell variables which are accessed by programs are therefore known as 'environment variables'[8]. You can see the full list of the environment variables that your shell currently recognizes by running:

> `$ printenv`

One of the most commonly used environment variables is `PATH`, it is a list of directories to search for executable names. The most basic way to run an executable is to explicitly type the full file name (including all the directory information) and run it. This is useful for simple shell scripts or programs that you don't use too often. However, when the program (an executable) is to be used a lot, specifying all those directories will become a significant burden. The `PATH` environment variable keeps the address of all the directories to be searched if directory information is not explicitly given[9]. When you don't have root access, you need to specify a directory for your self and add that to the `PATH` environment variable.

Adding your specified directory to the `PATH` environment variable each time you want to run your program is again very troubling and will not be much of an improvement compared to explicitly calling the executbale with directory information. So there are standard 'startup files' defined by your shell. The commands in these files are run each time you start your system (`/etc/profile` and all scripts in `/etc/profile.d/`), when you log in (`~/.bash_profile`) or on each invocation of the shell (the terminal, `~/.bashrc`)[10].

`HOME` is another commonly used environment variable, it is any user's (the one that is logged in) top directory. It is used so often that Bash has a special expansion for it: `~`, whenever you see file names starting with the tilde sign, it actually represents the value to the `HOME` environment variable. The standard directory where you can keep installed files for your own user is the `~/.local/`. You can use this directory as the top directory for installing all the programs (executables), libraries, manuals and shared data that you need.

Let's call the directory you have chosen with `USRDIR` since the standard is just a suggestion. Please replace it with any directory name you choose. To notify the build system of the program to install the files in this directory, you can add the following option to the configure script. When you subsequently run `$ make install` all the installable files will be put there.

> `$ ./configure --prefix=USRDIR`

---

[7] By running `$ export myvariable=a test value` instead of the simpler case in the text

[8] You can use shell variables for other actions too, for example to temporarily keep some names or run loops on some files.

[9] This is why in the sequence of commands in Section 1.1 [Quick start], page 1 only `$ ./configure` has directory information. By giving a specific directory (the current directory or `./`), we are explicitly telling the shell to look in the current directory for an executable named `configure` not in the directories listed in `PATH`.

[10] These directories are the standard in GNU Bash, other shells might have different startup files.

The `USRDIR/bin` directory is the place where the executables (or binary files) are installed. So you have to add that to your `PATH` environment variable by placing the following command in the `$HOME/.bashrc` file or any of the startup files discussed above. The directories listed in `$PATH` specify the locations that the system will check to find the executable name you have asked for. Each directory is separated by a colon (`:`). So through the command below you will concatenate your directory to the already existing list.

    export PATH=$PATH:USRDIR/bin

In case you install libraries (like the requirements of Gnuastro) with this method locally, you also have to notify the system to search for shared libraries in your installed directory. To do that add `USRDIR/lib` to your `LD_LIBRARY_PATH` environment variable similar to the example above for `PATH`. If you also want to access the Info and man pages documentations add the `USRDIR/share/info` and `USRDIR/share/man` to your `INFODIR` and `MANPATH` environment variables.

A final note is that order matters in the directories that are searched. In the example above, the new directory was added after the system specified directories. So if the program, library or manuals are found in the system wide directories, the user directory is no longer searched. If you want to search your local installation first, put the new directory before the already existing list like the example below.

    export PATH=USRDIR/bin:$PATH

This is good when a library for example CFITSIO is already present on the system but wasn't installed with the correct configuration flags discussed above. Since you can't re-install, with this order, the system will first find the one you installed with the correct configuration flags. However there are security problems, because all system wide programs and libraries can be replaced by non-secure versions if they also exist in `USRDIR`. So if you choose this order, be sure to keep it clean from executables with the same names as important system programs.

3.3.1.3 Executable names

At first sight, the names of the executables for each program might seem to be uncommonly long, for example **astnoisechisel** or **astimgcrop**. We could have chosen terse (and cryptic) names like most programs do. We chose this complete naming convention (something like the commands in TeX) so you don't have to spend too much time remembering what the name of a specific program was. Such complete names also enable you to easily search for the programs.

To facilitate typing the names in, we suggest using the shell auto-complete. With this facility you can find the executable you want very easily. It is very similar to file name completion in the shell. For example, simply by typing the letters bellow (where `[TAB]` stands for the Tab key on your keyboard)

    $ ast[TAB][TAB]

you will get the list of all the available executables that start with **ast** in your `PATH` environment variable directories. So, all the Gnuastro executables installed on your system will be listed. Typing the next letter for the specific program you want along with a Tab, will limit this list until you get to your desired program.

In case all of this does not convince you and you still want to type short names, some suggestions are given below. You should have in mind though, that if you are writing a shell script that you might want to pass on to others, it is best to use the standard name because other users might not have adopted the same customizations. The long names also serve as a form of documentation in such scripts. A similar reasoning can be given for option names in scripts: it is good practice to always use the long formats of the options in shell scripts, see Section 4.1.3 [Options], page 32.

The simplest solution is making a symbolic link to the actual executable. For example let's assume you want to type `ic` to run ImageCrop instead of `astimgcrop`. Assuming you installed Gnuastro executables in `/usr/local/bin` (default) you can do this simply by running the following command as root:

```
# ln -s /usr/local/bin/astimgcrop /usr/local/bin/ic
```

In case you update Gnuastro and a new version of ImageCrop is installed, the default executable name is the same, so your custom symbolic link still works.

The installed executable names can also be set using options to `$ ./configure`, see Section 3.3.1 [Configuring], page 24. GNU Autoconf (which configures Gnuastro for your particular system), allows the builder to change the name of programs with the three options `--program-prefix`, `--program-suffix` and `--program-transform-name`. The first two are for adding a fixed prefix or suffix to all the programs that will be installed. This will actually make all the names longer! You can use it to add versions of program names to the programs in order to simultaneously have two executable versions of a program.

The third configure option allows you to set the executable name at install time using the SED utility. SED is a very useful 'stream editor'. There are various resources on the internet to use it effectively. However, we should caution that using configure options will change the actual executable name of the installed program and on every re-install (an update for example), you have to also add this option to keep the old executable name updated. Also note that the documentation or configuration files do not change from their standard names either.

For example, let's assume that typing `ast` on every invocation of every program is really annoying you! You can remove this prefix from all the executables at configure time by adding this option:

```
$ ./configure --program-transform-name='s/ast/ /'
```

3.3.2 Tests

After successfully building (compiling) the programs with the `$ make` command you can check the installation before installing. To run the tests on your newly build utilities, run

```
$ make check
```

For every program some tests are designed to check some possible operations. Running the command above will run those tests and give you a final report. If everything is ok and you have built all the programs, all the tests should pass. In case any of the tests fail, please have a look at Section 3.3.4 [Known issues], page 29 and if that still doesn't fix your problem, look that the `./tests/test-suite.log` file to see if the source of the error is something particular to your system or more general. If you feel it is general, please contact us because it might be a bug. Note that the tests of some programs depend on the outputs of other program's tests, so if you have not installed them they might be skipped or fail. Prior to releasing every distribution all these tests are checked. If you have a reasonably modern terminal, the outputs of the successful tests will be colored green and the failed ones will be colored red.

These scripts can also act as a good set of examples for you to see how the programs are run. All the tests are in the `gnuastro-0.1/tests` directory. The tests for each program are shell scripts (ending with `.sh`) in a subdirectory of this directory with the same name as the program. See Section 10.8 [Test scripts], page 160 for more detailed information about these scripts incase you want to inspect them.

3.3.3 A4 print manual

The default print manual is provided in the letter paper size. If you would like to have the print version of this manual on paper and you are living in a country which uses A4, then you can rebuild the manual. The great thing about the GNU build system is that the manual source code which is in Texinfo is also distributed with the program source code, enabling you to do such customizations (hacking).

In order to change the paper size, you will need to have GNU Texinfo installed. For simplicity, let's assume `SRCdir` is equivalent to `DOWLD/gnuastro-0.1`. Open `SRCdir/doc/gnuastro.texi` with any text editor. This is the source file that created this manual. In the first few lines you will see this line:

    @c@afourpaper

In Texinfo, a line is commented with `@c`. Therefore, uncomment this line by deleting the first two characters such that it changes to:

    @afourpaper

Save the file and close it. You can now run

    $ make pdf

and the new PDF manual will be available in `SRCdir/doc/gnuastro.pdf`. By changing the `pdf` in `$ make pdf` to `ps` or `dvi` you can have the manual in those formats. Note that you can do this for any manual that is in Texinfo format, they might not have `@afourpaper` line, so you can add it close to the top of the Texinfo source file.

3.3.4 Known issues

Depending on your operating system and the version of the compiler you are using, you might confront some known problems during the configuration (`$ ./configure`), compilation (`$ make`) and tests (`$ make check`). Here, their solutions are discussed.

- `$ ./configure`: *Configure complains about not finding a library even though you have installed it.* The possible solution is based on how you installed the package:

    - From your distribution's package manager. Most probably this is because your distribution has separated the header files of a library from the library parts. Please also install the 'development' packages for those libraries too. Just add a `-dev` or `-devel` to the end of the package name and re-run the package manager. This will not happen if you install the libraries from source. When installed from source, the headers are also installed.

    - From source. Then your linker is not looking where you installed the library. If you followed the instructions in this chapter, all the libraries will be installed in `/usr/local/lib`. So you have to tell your linker to look in this directory. To do so, add `LDFLAGS=-L/usr/local/lib` to the Gnuastro configure script. If you want to use the libraries for your other programming projects, then export this environment variable similar to the case for `LD_LIBRARY_PATH` explained below.

- `$ make`: *Complains about an unknown function on a non-GNU based operating system.* In this case, please run `$ ./configure` with the `--enable-gnulibcheck` option to see if the problem is from the GNU Portability Library (Gnulib) not supporting your system or if there is a problem in Gnuastro, see Section 3.3.1.1 [GNU Astronomy Utilities configure options], page 24. If the problem is not in Gnulib and after all its tests you get the same complaint from `make`, then please contact us at `bug-gnuastro@gnu.org`. The cause is probably that a function that we have used is not supported by your operating system and we didn't included it along with the source tar ball. If the function is available in Gnulib, it can be fixed immediately.

- `$ make`: *Can't find the headers (.h files) of libraries installed from source.* Similar to the case for `LDFLAGS` (above), your compiler is not looking in the right place, add `CPPFLAGS=-I/usr/local/include` to `./configure` to re-configure Gnuastro, then re-run make.

- `$ make check`: *Only one (the first) test passes, all the rest fail.* It is highly likely that your distribution doesn't look into the `/usr/local/lib` directory when searching for shared libraries. To make sure it is added to the list of directories, run the following command and restart your terminal: (you can ignore the \ and extra space if you type it, it is only necessary if you copy and paste). See Section 3.3.1.2 [Installation directory], page 25 for more details.

```
echo 'export LD_LIBRARY_PATH=$LD_LIBRARY_PATH:/usr/local/lib' \
     >> ~/.bashrc
```

- `$ make check`: *The tests relying on external programs (for example `fitstopdf.sh` fail.)* This is probably due to the fact that the version number of the external programs is too old for the tests we have preformed. Please update the program to a more recent version. For example to create a PDF image, you will need GPL Ghostscript, but older versions do not work, we have successfully tested it on version 9.15. Older versions might cause a failure in the test result.

- `$ make pdf`: *The PDF manual cannot be made.* To make a PDF manual, you need to have the GNU Texinfo program (like any program, the more recent the better). A working TEX program is also necessary, which you can get from Tex Live[11].

If your problem was not listed above, please file a bug report (Section 1.7 [Report a bug], page 7).

---

[11] https://www.tug.org/texlive/

4 Common behavior

There are some facts that are common to all the programs in Gnuastro which are mainly to do with user interaction. In this chapter these aspects are discussed. The most basic are the command line options which are common in all the programs for a unified user experience. All Gnuastro programs can use configuration files so you don't have to specify all the parameters on the command line each time you run a program. The manner of setting, checking and using the these files at various levels are also explained. Finally we discuss how you can get immediate and distraction-free (without taking your hands off the keyboard!) help on the command line.

4.1 Command line

All the programs in GNU Astronomy Utilities are customized through the standard GNU style command line options. First a general outline of how to make best use of these options is discussed and finally the options that are common to all the programs in Gnuastro are listed.

Your full command line text is passed onto the shell as a string of characters. That string is then broken up into separate 'words' by any 'metacharacters' (like space, tab, |, > or ;) that might exist in the text. See Section "Definitions" in *the Bash manual*, for the complete list of meta-characters and other Bash definitions. See Section "Shell Operation" in *the Bash manual*, for a short summary of the steps the shell takes before passing the commands to the program you called.

4.1.1 Arguments and options

On the command line, the first thing you enter is the name of the program you want to run. After that you can specify two types of input: *arguments* and *options*. Arguments are those tokens that are not preceded by any hyphens (-), the program is suppose to understand what they are without any help from the user.

Arguments can be both mandatory and optional and since there is no help from you, their order might also matter (for example in `cp` which is used for copying). The outputs of `--usage` and `--help` shows which arguments are optional and which are mandatory, see Section 4.6.1 [--usage], page 44. As their name suggests, *options* are only optional and most of the time you don't have to worry about what order you specify them in.

In case your arguments or option values contain any of the shell's meta-characters, you have to quote them. If there is only one such character, you can use a backslash (\) before it. If there are multiple, it might be easier to simply put your whole argument or option value inside of double quotes ("). In such cases, everything inside the double quotes will be seen as one 'word'.

For example let's say you want to specify the Header data unit (HDU) of your FITS file using a complex expression like `3; images(exposure > 100)`. If you simply add these after the `--hdu` (`-h`) option, the programs in Gnuastro will read the value to the HDU option as `3` and run. Then, Bash will attempt to run a separate command `images(exposure > 100)` and complain about a syntax error. This is because the semicolon (;) is an 'end of command' character in Bash. To solve this problem you can simply put double quotes around the whole string you want to pass as seen below:

```
$ astimgcrop --hdu="3; images(exposure > 100)" FITSimage.fits
```

Alternatively you can put a \ before every metacharacter in this string, but probably you will agree with us that the double quotes are much more easier, elegant and readable.

4.1.2 Arguments

In GNU Astronomy Utilities, the names of the input data files and ASCII tables are mostly specified as arguments, you can generally specify them in any order unless otherwise stated for a particular program. Everything particular about how a program treats arguments, is explained under the "Invoking ProgramName" section for that program.

Generally, if there is a standard file name extension for a particular format, that filename extension is used to separate the kinds of arguments. The list below shows what astronomical data formats are recognized based on their file name endings. If the program doesn't accept any other data format, any other argument that doesn't end with the specified extentions below is considered to be a text file (usually catalogs). For example Section 5.2 [ConvertType], page 51 accepts other data formats.

- .fits: The standard file name ending of a FITS image.
- .fits.Z: A FITS image compressed with compress.
- .fits.gz: A FITS image compressed with GNU zip (gzip).
- .fits.fz: A FITS image compressed with fpack.
- .imh: IRAF format image file.

Through out this manual and in the command line outputs, whenever we want to generalize all such astronomical data formats in a text place holder, we will use ASTRdata, we will assume that the extension is also part of this name. Any file ending with these names is directly passed on to CFITSIO to read. Therefore you don't necessarily have to have these files on your computer, they can also be located on an FTP or HTTP server too, see the CFITSIO manual for more information.

CFITSIO has its own error reporting techniques, if your input file(s) cannot be opened, or read, those errors will be printed prior to the final error by Gnuastro.

4.1.3 Options

Command line options allow configuring the behaviour of a program in all GNU/Linux applications for each particular execution. Most options can be called in two ways: *short* or *long* a small number of options in some programs only have the latter type. In the list of options provided in Section 4.1.4 [Common options], page 34 or those for each program, both formats are shown for those which support both. First the short is shown then the long. Short options are only one character and only have one hyphen (for example -h) while long options have two hyphens an can have many characters (for example --hdu).

Usually, the short options are for when you are writing on the command line and want to save keystrokes and time. The long options are good for shell scripts, where you don't usually have a rush and they provide a level of documentation, since they are less cryptic. Usually after a few months of not running a program, the short options will be forgotten and reading your previously written script will not be easy.

Some options need to be given a value if they are called and some don't. You can think of the latter type of options as on/off options. These two types of options can be distinguished using the output of the `--help` and `--usage` options, which are common to all GNU software, see Section 4.6 [Getting help], page 43. The following convention is used for the formats of the values in Gnuastro:

INT   The value is read as an integer. If a float or a string is provided the program will warn you and abort. In most cases, integers are used for counting variables, so if they are negative the program will also abort.

4or8   Either the value 4 or 8, any other integer will give a warning and abort.

FLT   The value is read as a float. There are generally two types, depending on the context. If they are for fractions, they will have to be less than or equal to unity.

STR   The value is read as a string of characters (for example a file name) or other particular settings like a HDU name, see below.

To specify a value in the short format, simply put the value after the option. Note that since the short options are only one character long, you don't have to type anything between the option and its value. For the long option you either need white space or an = sign, for example -h2, -h 2, --hdu 2 or --hdu=2 are all equivalent.

The short format of on/off options (those that don't need values) can be concatenated for example these two hypothetical sequences of options are equivalent: -a -b -c4 and -abc4. As an example, consider the following command to run ImageCrop:

```
$ astimgcrop -Dr3 --wwidth 3 catalog.txt --deccol=4 ASTRdata
```

The $ is the shell prompt, `astimgcrop` is the program name. There are two arguments (`catalog.txt` and `ASTRdata`) and four options, two of them given in short format (-D, -r) and two in long format (--width and --deccol). Three of them require a value and one (-D) is an on/off option.

If an abbreviation is unique between all the options of a program, the long option names can be abbreviated. For example, instead of typing `--printparams`, typing `--print` or maybe even `--pri` will be enough, if there are conflicts, the program will warn you and show you the alternatives. Finally, if you want the argument parser to stop parsing arguments beyond a certain point, you can use two dashes: `--`. No text on the command line beyond these two dashes will be parsed.

If an option with a value is repeated or called more than once, the value of the last time it was called will be assigned to it. This very useful in complicated sitations, for example in scripts. Let's say you want to make a small modification to one option value. You can simply type the option with a new value in the end of the command and see how the script works. If you are satisfied with the change, you can remove the original option. If the change wasn't satsifactory, you can remove the one you just added and not worry about saving the original value. Without this capability, you would have to memorize or save the original value somewhere else, run the command and then change the value again which is not at all convenient and is potentially cause lots of bugs.

When you don't call an option that requires a value, all the programs in Gnuastro will check configuration files to find a value for that parameter. To learn more about how folder, user and system wide configuration files can be set, please see Section 4.2 [Configuration files], page 37. Another factor that is particular to Gnuastro is that it will check the value you have given for each option to see if it is reasonable. For example you might mistakenly give a negative, float or string value for a FITS image extension or column number. As another example, you might give a value larger than unity for an option that only accepts fractions (which are always less than unity and positive).

> **CAUTION:** In specifying a file address, if you want to use the shell's tilde expansion (~) to specify your home directory, leave at least one space between the option name and your value. For example use -o ~/test, --output ~/test or --output= ~/test. Calling them with -o~/test or --output=~/test will disable shell expansion.

> **CAUTION:** If you forget to specify a value for an option which requires one, and that option is the last one, Gnuastro will warn you. But if it is in the middle of the command, it will take the text of the next option or argument as the value which can cause undefined behaviour.

> **NOTE:** All counting in Gnuastro starts from 0 not 1. So for example the first FITS image extension or column in a table are noted by 0, not 1. This is the standard in C and all languages that are based on it (for example C++, Java and Python).

4.1.4 Common options

To facilitate the job of the users and developers, all the programs in Gnuastro share some basic command line options for the same operations where they are relevant. The list of options is provided below. It is noteworthy that these similar options are hard-wired into the programming of all of Gnuastro programs using GNU C Library's argument parser merging ability.

For some programs, some of the options, might be irrelevant for example MakeProfiles creates FITS images based on a given catalog. Therefore no input images (and thus HDUs) are necessary for it. In such cases, the option is still listed and if a value is given for it, it is completely ignored.

4.1.4.1 Input/Output options

These options are to do with the input and outputs of the various programs.

-h

--hdu   (=STR) The number or name of the desired Header Data Unit or HDU in the input FITS image or images. A FITS file can store multiple HDUs or extensions, each with either an image or a table or nothing at all (only a header). Note that counting of the extensions starts from 0(zero), not 1(one). When specifying the name, case is not important so IMAGE, image or ImAgE are equivalent.

       A # is appended to the string you specify for the HDU[1] and the result is put in square brackets and appended to the FITS file name before calling CFITSIO

---

[1] With the # character, CFITSIO will only read the desired HDU into your memory, not all the existing HDUs in the fits file.

to read the contents of the HDU for all the programs in Gnuastro. CFITSIO has many capabilities to help you find the extension you want, far beyond the simple extension number and name. See CFITSIO manual's "HDU Location Specification" section for a very complete explanation with several examples.

`-o`
`--output`   (=STR) The name of the output file or directory. With this option the automatic output names explained in Section 4.5 [Automatic output], page 43 are ignored.

`-D`
`--dontdelete`

By default, if the output file already exists, it will be silently replaced with the output of this run of all Gnuastro programs. By calling this option, if the output file already exists, the programs will warn you and abort.

`-K`
`--keepinputdir`

In automatic output names, don't remove the directory information of the input file names. As explained in Section 4.5 [Automatic output], page 43, if no output name is specified, then the output name will be made in the existing directory based on your input. If you call this option, the directory information of the input will be kept and the output will be in the same directory as the input. Note that his is only relevant if you are running the program from another directory!

4.1.4.2 Operating modes

Another group of options that are common to all the programs in Gnuastro are those to do with the general operation of the programs. The explanation for those that are not only limited to Gnuastro but can be called in all GNU programs start with (GNU option).

`--`   (GNU option) Stop parsing the command line. This option can be useful in scripts or when using the shell history. Suppose you have a long list of options, and want to see if removing some of them (and using the default values) can give a better result. If the ones you want to remove are the last ones on the command line, you don't have to delete them, you can just add `--` before them and if you don't get what you want, you can remove the `--` and get the same initial result.

`--usage`   (GNU option) Only print the options and arguments. This is very useful for when you know the what the options do, you have just forgot their names. See Section 4.6.1 [--usage], page 44.

`-?`
`--help`   (GNU option) Print all options and an explanation. Adding this option will print all the options in their short and long formats, also displaying which ones need a value if they are called (with an = after the long format). A short explanation is also given for what the option is for. The program will quit immediately after the message is printed and will not do any form of processing. See Section 4.6.2 [--help], page 44.

-V

--version

(GNU option) Print a short message, showing the full name, version, copyright information and program authors. On the first line it will print the official name (not executable name) and version number of the program. It will also print the version of the Gnuastro that the program was built with. Following this is a blank line and a copyright information. The program will not run.

-q

--quiet   Don't report steps. All the programs in Gnuastro that have multiple major steps will report their steps for you to follow while they are operating. If you do not want to see these reports, you can call this option and only error messages will be printed if the program is aborted. If the steps are done very fast (depending on the properties of your input) disabling these reports will also decrease running time.

--cite   Print the BibTEX entry for Gnuastro and the particular program (if that program comes with a separate paper) and abort. Citations are vital for the continued work on Gnuastro. Gnuastro started and is continued based on separate research projects. So if you find any of the tools offered in Gnuastro to be useful in your research, please use the output of this command to cite the program and Gnuastro in your research paper. Thank you.

GNU Astronomy Utilities is still new, there is no separate paper only devoted to Gnuastro yet. Therefore currently the paper to cite for Gnuastro is the paper for NoiseChisel which is the first published paper introducing Gnuastro to the astronomical community. Upon reaching a certain point, a paper completely devoted to Gnuastro will be published, see Section 1.5.1 [GNU Astronomy Utilities 1.0], page 5.

-P

--printparams

Print the final values used for all the parameters and abort. See Section 4.4 [Final parameter values, reproduce previous results], page 42 for more details.

-S

--setdirconf

Update the current directory configuration file from the given command line options and quit, see Section 4.2 [Configuration files], page 37. The values of your options are added to the configuration file in the current directory. If the configuration file or folder doesn't exist, it will be created. If it exists but has different values for those options, they will be given the new values. In any case, the program will not run, but the contents of its updated configuration file are printed for you to inspect.

This is the recommended method to fill the configuration file for all future calls to one of the Gnuastro programs in a folder. It will internally check if your values are in the correct range and type and save them according to the configuration file format, see Section 4.2.1 [Configuration file format], page 38.

> When this option is called, the otherwise mandatory arguments, for example
> input image or catalog file(s), are no longer mandatory (since the program will
> not run).

`-U`

`--setusrconf`

> Update the user configuration file from the command line options and quit. See
> explanation under `--setdirconf` for more details.

`--onlydirconf`

> Only read the current (local) directory configuration file and ignore the rest of
> the configuration files, see Section 4.2.2 [Configuration file precedence], page 38
> and Section 4.2.3 [Current directory and User wide], page 39. This can be
> very useful when you want your results to be exactly reproducible. All the
> configuration files can be put in the hidden `./.gnuastro/` directory in the
> current directory, or the hidden directory can be a symbolic link to the directory
> containing the configuration files. Then with this option you can ensure that
> no other configuration file is read. So if your local configuration file lacks
> some parameters, which ever Gnuastro utility you are using will will warn you
> and abort, enabling you to exactly set all the necessary parameters without
> unknowningly relying on some user or system wide option values.

> `onlydirconf` can also be used in the configuration files (with a value of 0 or
> 1), see Section 4.2.1 [Configuration file format], page 38. If it is present in the
> local configuration file, other configuration files will not be read. In the other
> configuration files, it is irrelevant.

`--onlyversion`

> (=STR) Only run the program if its version is equal with the string of characters
> given to this option. Note that it is not compared as a number, but as a string
> of characters, so 0, or 0.0 and 0.00 are different. This is useful if you want
> your results to be exactly reproducible and not mistakenly run with an updated
> or older version of the program.

`--nolog` For programs which generate Log files, if this option is called, no Log file will
> be generated.

`-N`

`--numthreads`

> (=INT) Set the number of CPU threads to use. See Section 4.3 [Threads in
> GNU Astronomy Utilities], page 40.

4.2 Configuration files

Each program needs a certain number of parameters to run. Supplying all the necessary
parameters each time you run the program is very frustrating and prone to errors. Therefore
all the programs read the values for the necessary options you have not given in the command
line from one of several plain text files (which you can view and edit with any text editor).
These files are known as configuration files and are usually kept in a directory named `etc/`
according to the file system hierarchy standard[2].

---

[2] `http://en.wikipedia.org/wiki/Filesystem_Hierarchy_Standard`

The thing to have in mind is that none of the programs in Gnuastro keep any internal default value. All the values must either be stored in one of the configuration files or explicitly called in the command line. In case the necessary parameters are not given through any of these methods, the program will list the missing necessary parameters and abort. The only exception to this is `--numthreads`, whose default value is set at `$ ./configure` time internally, see Section 4.3 [Threads in GNU Astronomy Utilities], page 40. Of course, you can still provide a default value for the number of threads at any of the levels below, but if you don't, the program will not abort. Also note that through automatic output name genertion, the value to the `--output` option is also not mandatory on the command line or in the configuration files for all programs which don't rely on that value as an input[3], see Section 4.5 [Automatic output], page 43.

4.2.1 Configuration file format

The configuration files for each program have the standard program executable name with a `.conf` suffix. When you download the source code, you can find them in the same directory as the source code of each program, see Section 10.7 [Program source], page 154.

Any line in the configuration file whose first non-white character is a `#` is considered to be a comment and is ignored. The same goes for an empty line. The name of the parameter is the same as the long format of the command line option for that parameter. The parameter name and parameter value have to be separated by any number of 'white-space' characters: space, tab or vertical tab. By default several space characters are used. If the value of an option has space characters (most commonly for the `hdu` option), then double quotes can be used to specify the full value.

Any text after the first two words (separated by the above delimiters) in a line is ignored. If it is an option without a value in the `--help` output (on/off option), then the value should be `1` if it is to be 'on' and `0` otherwise. If an option is not recognized in the configuration file, the name of the file and unrecognized option will be reported and the program will abort. If a parameter is repeated more more than once in the configuration files and it is not set on the command line, then only the first value will be used, the rest will be ignored.

You can build or edit any of the directories and the configuration files your self using any text editor. However, it is recommended to use the `--setdirconf` and `--setusrconf` options to set default values for the current directory or this user, see Section 4.1.4.2 [Operating modes], page 35. With these options, the values you give will be checked as explained in Section 4.1.3 [Options], page 32 before writing in the current directory's configuration file. They will also print a set of commented lines guiding the reader and will also classify the options based on their context and write them in their logical order to be more understandable.

4.2.2 Configuration file precedence

The parameter values in all the programs of Gnuastro will be filled in the following order. Such that if a parameter is assigned a value in an earlier step, any value for that parameter in a later step will be ignored.

---

[3] One example of a program which uses the value given to `--output` as an input is ConvertType, this value specifies the type of the output through the value to `--output`, see Section 5.2.3 [Invoking ConvertType], page 54.

1. Command line options, for this particular execution.

2. Current directory, for all executions in the directory from which any of the utilities is run (`./.gnuastro/`).

3. The user's home directory, for all the executions of a particular user: (`$HOME/.local/etc/`, see below). It is only read if `--onlydirconf` is not called, see Section 4.1.4.2 [Operating modes], page 35.

4. In a system wide directory for any user on that computer (`prefix/etc/`, see Section 3.3.1.2 [Installation directory], page 25 for the value of `prefix`). It is only read if `--onlydirconf` is not called, see Section 4.1.4.2 [Operating modes], page 35.

The basic idea behind setting this progressive state of checking for parameter values is that separate users of a computer or separate folders in a user's file system might need different values for some parameters and the same values for others. For example raw telescope images usually have their main image extension in the second FITS extension, while processed FITS images usually only have one extension. If your system wide default input extension is 0 (the first), then when you want to work with the former group of data you have to explicitly mention it to the programs every time. With this progressive state of default values to check, you can set different default values for the different directories that you would like to run Gnuastro in for your different purposes, so you won't have to worry about this issue any more.

4.2.3 Current directory and User wide

For the current (local) and user-wide directories, the configuration files are stored in the hidden sub-directories named `./.gnuastro/` and `HOME/.local/etc/` respectively. Unless you have changed it, the `HOME` environment variable should point to your home directory. You can check it by running `$ echo $HOME`. Each time you run any of the programs in Gnuastro, this environment variable is read and placed in the above address. So if you suddenly see that your home configuration files are not being read, probably you (or some other program) has changed the value of this environment variable.

Although it might cause confusions like above, this dependence on the `HOME` environment variable enables you to temporarily use a different directory as your home directory. This can come in handy in complicated situations. To set the user or current directory configuration files based on your command line input, you can use the `--setdirconf` or `--setusrconf`, see Section 4.1.4.2 [Operating modes], page 35

4.2.4 System wide

When Gnuastro is installed, the configuration files that are shipped with the distribution are copied into the (possibly system wide) `prefix/etc/` directory. See Section 3.3.1 [Configuring], page 24 for more details on `prefix` (by default it is: `/usr/local`). This directory is the final place (with the lowest priority) that the programs in Gnuastro will check to retrieve parameter values.

If you remove a parameter and its value from the files in this system wide directory, you either have to specify it in more immediate configuration files or set it each time in the command line. Recall that none of the programs in Gnuastro keep any internal default values and will abort if they don't find a value for the necessary parameters (except the number of threads). So even though you might never use a parameter, it still has to be at least available in this system-wide configuration file.

In case you install Gnuastro from your distribution's repositories, `prefix` will either be set to / (the root directory) or `/usr`, so you can find the system wide configuration variables in `/etc/` or `/usr/etc/`. The prefix of `/usr/local/` is conventionally used for programs you install from source by your self.

4.3 Threads in GNU Astronomy Utilities

Some of the programs benefit significantly when you use all the threads your computer's CPU has to offer to your operating system. GNU Astronomy Utilities uses the POSIX threads library (pthreads) for spinning off threads when the user asks for it. The number of threads available to your operating system is usually double the number of physical (hardware) cores in your CPU.

You can find the number of threads available to your system with the command $ `nproc`, which is part of GNU Coreutils and is most probably already available on your GNU/Linux system. If not specified as an option at configure time, Gnuastro finds the number of threads available to your system (and reports it along with all those other things it checks!). It is saved internally for all the programs to use by default. To specify the number of threads at configure time, use the `--with-numthreads` option, see Section 3.3.1.1 [GNU Astronomy Utilities configure options], page 24. In case your system does not have GNU Coreutils, currently the only way to proceed is to manually specify the number of threads through this option.

The number of threads is the only parameter in Gnuastro which is stored internally at configure time. The implication is that the only option with a value that doesn't have to be in any of the configuration files is this, see Section 4.2 [Configuration files], page 37. Note that if you do specify it, the value you provided in the most immediate configuration file will be used, not the internal value.

4.3.1 A note on threads

Spinning off threads internally is not necessarily always the most efficient way to run an application. Creating a new thread isn't a cheap operation for the operating system. It is most useful when the input data are fixed and you want the same operation to be done on parts of it. For example one input image to ImageCrop and multiple crops from various parts of it. In this fashion, the image is loaded into memory once, all the crops are divided between the number of threads internally and each thread cuts out those parts which are assigned to it from the same image. On the other hand, if you have multiple images and you want to crop the same region out of all of them, it is much more efficient to set `--numthreads=1` (so no threads spin off) and run ImageCrop multiple times simultaneously, see Section 4.3.2 [How to run simultaneous operations], page 41.

You can check the boost in speed by first running a program on one of the data sets with the maximum number of threads and another time (with everything else the same) and only using one thread. You will notice that the wall-clock time (reported by most programs at their end) in the former is longer than the latter divided by number of physical CPU cores available to your operating system. Asymptotically these two can be equal (most of the time they aren't). So limiting the programs to use only one thread and running them independently on the number of available threads will be more efficient.

Note that the operating system keeps a cache of recently processed data, so usually, the second time you process an identical dataset (independent of the number of threads used), you will get faster results. In order to make an unbiased comparison, you have to first clean the system's cache with the following command between the two runs.

```
$ sync; echo 3 | sudo tee /proc/sys/vm/drop_caches
```

SUMMARY: Should I use multiple threads? Depends:

- If you only have **one** data set (image in most cases!), then yes, the more threads you use (with a maximum of the number of threads available to your OS) the faster you will get your results.

- If you want to run the same operation on **multiple** data sets, it is best to set the number of threads to 1 and use GNU Parallel as explained above.

4.3.2 How to run simultaneous operations

There are two approaches to simultaneously execute a program: using GNU Parallel or Make (GNU Make is the most common implementation). The first is very useful when you only want to do one job multiple times and want to get back to your work without actually keeping the command you ran. The second is usually for (very) complicated processes, with lots of dependancies between the different products (for example a data-production pipeline).

GNU Parallel

> When you only want to run multiple instances of a command on different threads and get on with the rest of your work, the best method is to use GNU parallel. Surprisingly GNU Parallel is one of the few GNU packages that has no Info documentation but only a Man page, see Section 4.6.4 [Info], page 46. So to see the documentation after installing it please run
>
> ```
> $ man parallel
> ```
>
> As an example, let's assume we want to crop a region fixed on the pixels (500, 600) with the default width from all the FITS images in the ./data directory ending with sci.fits to the current directory. To do this, you can run:
>
> ```
> $ parallel astimgcrop --numthreads=1 --xc=500 --yc=600 ::: \
>     ./data/*sci.fits
> ```
>
> GNU Parallel can help in many more conditions, this is one of the simplest, see the man page for lots of other examples. For absolute beginners: the backslash (\) is only a line breaker to fit nicely in the page. If you type the whole command in one line, you should remove it.

Make

> Make is a utility built for specifying "targets", "prerequisites" and "receipts". It allows you to define very complicated dependancy structures for complicated processes that commonly start off with a large list of inputs and builds them based on the dependancies you define. GNU Make[4] is the most common im-

---

[4] https://www.gnu.org/software/make/

plementation which (like nearly all GNU programs comes with a wonderful manual[5]). It is very basic and short (the most important part is about 100 pages).

Make has the `--jobs` (`-j`) which allows you to specify the maximum number of jobs that can be done simultaneously. For example a common 4 physical core CPU usually has 8 processing threads. So you can run:

```
$ make -j8
```

Once the dependancy tree for your processes is built, Make will run the independent targets simultaneously.

4.4 Final parameter values, reproduce previous results

The input parameters can be specified in many places, either on the command line or in at least one of several configuration files, see Section 4.2 [Configuration files], page 37. Therefore, it often happens that before running a program on a certain data set, you want to see the values for the parameters that the program will use after it has read your command line options and all the configuration files in their correct order. You might also want to save the list with the output so you can reproduce the same results at a later time, this is very important when you want to use your results in a report or paper.

If you call the `--printparams` option, all Gnuastro programs will read your command line parameters and all the configuration files. If there is no problem (like a missing parameter or a value in the wrong format) and immediately before actually running, the programs will print the full list of parameter names and values sorted and grouped by context and quit. They will also report their version number, the date they were configured on your system and the time they were reported.

As an example, you can give your full command line options and even the input and output file names and finally just add `-P` to check if all the parameters are finely set. If everything is ok, you can just run the same command (easily retrieved from the bash history, with the top arrow key) and simply remove the last two characters that showed this option.

Since no program will actually start its processing when this option is called, the otherwise mandatory arguments for each program (for example input image or catalog files) are no longer required when you call this option.

In case you want to store the list of parameters for later reproduction of the same results, you can do so with the GNU Bash re-direction tool. For example after you have produced the results you want to store, you can save all the parameters that were used in a file named `parameters.txt` in the following manner. Using shell history you can retrieve the last command you entered and simply add `-P > parameters.txt` to it, for example:

```
$ astimgcrop --racol=2 --deccol=3 IN.fits cat.txt -P > parameters.txt
```

All the parameters along with the extra data explained before will be stored in the plain text `parameters.txt` file through the shell's redirection mechanism (`>`). The output of `--printparams` conforms with the configuration file formats[6]. By taking the following steps, you can use this file as a configuration file to reproduce your results at a later time.

---

[5] https://www.gnu.org/software/make/manual/

[6] They are both written by the same function.

1. Set the file name based on the standard configuration file names, see Section 4.2.1 [Configuration file format], page 38.

2. Later on (when ever you want to re-produce your results), copy the file in the ./.gnuastro/ directory of your current directory.

In this manner, this file will be read as a current directory configuration file and since all the parameters are defined in it, no other configuration file value will be used.

4.5 Automatic output

All the programs in Gnuastro are designed such that specifying an output file or directory (based on the program context) is optional. The outputs of most programs are automatically found based on the input and what the program does. For example when you are converting a FITS image named `FITSimage.fits` to a JPEG image, the JPEG image will be saved in `FITSimage.jpg`.

Another very important part of the automatic output generation is that all the directory information of the input file name is stripped off of it. This feature can be disabled with the `--keepinputdir` option, see Section 4.1.4 [Common options], page 34. It is the default because astronomical data are usually very large and organized specially with special file names. In some cases, the user might not have write permissions in those directories. In fact, even if the data is stored on your own computer, it is advised to only grant write permissions to the super user or root. This way, you won't accidentally delete or modify your valuable data!

Let's assume that we are working on a report and want to process the FITS images from two projects (ABC and DEF), which are stored in the sub-directories named `ABCproject/` and `DEFproject/` of our top data directory (`/mnt/data`). The following shell commands show how one image from the former is first converted to a JPEG image through Convert-Type and then the objects from an image in the latter project are detected using NoiseChisel. The text after the # sign are comments (not typed!).

```
$ pwd                                      # Current location
/home/usrname/research/report
$ ls                                       # List directory contents
ABC01.jpg
$ ls /mnt/data/ABCproject                            # Archive 1
ABC01.fits ABC02.fits ABC03.fits
$ ls /mnt/data/DEFproject                            # Archive 2
DEF01.fits DEF02.fits DEF03.fits
$ astconvertt /mnt/data/ABCproject/ABC02.fits --output=jpg    # Prog 1
$ ls
ABC01.jpg ABC02.jpg
$ astnoisechisel /mnt/data/DEFproject/DEF01.fits              # Prog 2
$ ls
ABC01.jpg ABC02.jpg DEF01_labeled.fits
```

4.6 Getting help

Probably the first time you read this manual, it is either in the PDF or HTML formats. These two formats are very convenient for when you are not actually working, but when you are only reading. Later on, when you start to use the programs and you are deep in the middle of your work, some of the details will inevitably be forgotten. Going to find the PDF file (printed or digital) or the HTML webpage is a major distraction.

GNU software have a very unique set of tools for aiding your memory on the command line, where you are working, depending how much of it you need to remember. In the past, such command line help was known as "online" help, because they were literally provided to you 'on' the command 'line'. However, nowadays the word "online" refers to something on the internet, so that term will not be used. With this type of help, you can resume your exciting research without taking your hands off the keyboard.

Another major advantage of such command line based help routines is that they are installed with the software in your computer, therefore they are always in sync with the executable you are actually running. Three of them are actually part of the executable. You don't have to worry about the version of the manual or program. If you rely on external help (a PDF in your personal print or digital archive or HTML from the official webpage) you have to check to see if their versions fit with your installed program.

If you only need to remember the short or long names of the options, --usage is advised. If it is what the options do, then --help is a great tool. Man pages are also provided for those who are use to this older system of documentation. This full manual is also available to you on the command line in Info format. If none of these seems to resolve the problems, there is a mailing list which enables you to get in touch with experienced Gnuastro users. In the subsections below each of these methods are reviewed.

4.6.1 --usage

If you give this option, the program will not run. It will only print a very concise message showing the options and arguments. Everything within square brackets ([]) is optional. For example here are the first and last two lines of ImageCrop's --usage is shown:

```
$ astimgcrop --usage
Usage: astimgcrop [-Do?IPqSVW] [-d INT] [-h INT] [-r INT] [-w INT]
            [-x INT] [-y INT] [-c INT] [-p STR] [-N INT] [--deccol=INT]
            ....
            [--setusrconf] [--usage] [--version] [--wcsmode]
            [ASCIIcatalog] FITSimage(s).fits
```

There are no explanations on the options, just their short and long names shown separately. After the program name, the short format of all the options that don't require a value (on/off options) is displayed. Those that do require a value then follow in separate brackets, each displaying the format of the input they want, see Section 4.1.3 [Options], page 32. Since all options are optional, they are shown in square brackets, but arguments can also be optional. For example in this example, a catalog name is optional and is only required in some modes. This is a standard method of displaying optional arguments for all GNU software.

4.6.2 --help

If the command line includes this option, the program will not be run. It will print a complete list of all available options along with a short explanation. The options are also grouped by their context. Within each context, the options are sorted alphabetically. Since the options are shown in detail afterwards, the first line of the `--help` output shows the arguments and if they are optional or not, similar to Section 4.6.1 [--usage], page 44.

In the `--help` output of all programs in Gnuastro, the options for each program are classified based on context. The first two contexts are always options to do with the input and output respectively. For example input image extensions or supplementary input files for the inputs. The last class of options is also fixed in all of Gnuastro, it shows operating mode options. Most of these options are already explained in Section 4.1.4.2 [Operating modes], page 35.

The help message will sometimes be longer than the vertical size of your terminal. If you are using a graphical user interface terminal emulator, you can scroll the terminal with your mouse, but we promised no mice distractions! So here are some suggestions:

- `Shift + PageUP` to scroll up and `Shift + PageDown` to scroll down. For most help output this should be enough. The problem is that it is limited by the number of lines that your terminal keeps in memory and that you can't scroll by lines, only by whole screens.

- Pipe to `less`. A pipe is a form of shell re-direction. The `less` tool in Unix-like systems was made exactly for such outputs of any length. You can pipe (`|`) the output of any program that is longer than the screen to it and then you can scroll through (up and down) with its many tools. For example:

      $ astnoisechisel --help | less

  Once you have gone through the text, you can quit `less` by pressing the `q` key.

- Redirect to a file. This is a less convenient way, because you will then have to open the file in a text editor! You can do this with the shell redirection tool (`>`):

      $ astnoisechisel --help > filename.txt

In case you have a special keyword you are looking for in the help, you don't have to go through the full list. GNU Grep is made for this job. For example if you only want the list of options whose `--help` output contains the word "axis" in ImageCrop, you can run the following command:

      $ astimgcrop --help | grep axis

If the output of this option does not fit nicely within the confines of your terminal, GNU does enable you to customize its output through the environment variable `ARGP_HELP_FMT`, you can set various parameters which specify the formatting of the help messages. For example if your terminals are wider than 70 spaces (say 100) and you feel there is too much empty space between the long options and the short explanation, you can change these formats by giving values to this environment variable before running the program with the `--help` output. You can define this environment variable in this manner:

      $ export ARGP_HELP_FMT=rmargin=100,opt-doc-col=20

This will affect all GNU programs using GNU C Library's `argp.h` facilities as long as the environment variable is in memory. You can see the full list of these formatting parameters in the "Argp User Customization" part of the GNU C Library manual. If you are more comfortable to read the `--help` outputs of all GNU software in your customized format, you can add your customizations (similar to the line above, without the `$` sign) to your `~/.bashrc` file. This is a standard option for all GNU software.

4.6.3 Man pages

Man pages were the Unix method of providing command line documentation to a program. With GNU Info, see Section 4.6.4 [Info], page 46 the usage of this method of documentation is highly discouraged. This is because Info provides a much more easier to navigate and read environment.

However, some operating systems require a man page for packages that are installed and some people are still used to this method of command line help. So the programs in Gnuastro also have Man pages which are automatically generated from the outputs of `--version` and `--help` using the GNU help2man program. So if you run

```
$ man programname
```

You will be provided with a man page listing the options in the standard manner.

4.6.4 Info

Info is the standard documentation format for all GNU software. It is a very useful command line document viewing format, fully equipped with links between the various pages and menus and search capabilities. As explained before, the best thing about it is that it is available for you the moment you need to refresh your memory on any command line tool in the middle of your work without having to take your hands off the keyboard. This complete manual is available in Info format and can be accessed from anywhere on the command line.

To open the Info format of any installed programs or library on your system which has an Info format manual, you can simply run the command below (change `executablename` to the executable name of the program or library):

```
$ info executablename
```

In case you are not already familiar with it, run `$ info info`. It does a fantastic job in explaining all its capabilities its self. It is very short and you will become sufficiently fluent in about half an hour. Since all GNU software documentation is also provided in Info, your whole GNU/Linux life will significantly improve.

Once you've become an efficient navigator in Info, you can go to any part of this manual or any other GNU software or library manual, no matter how long it is, in a matter of seconds. It also blends nicely with GNU Emacs (a text editor) and you can search manuals while you are writing your document or programs without taking your hands off the keyboard, this is most useful for libraries like the GNU C library. To be able to access all the Info manuals installed in your GNU/Linux within Emacs, type `Ctrl-H + i`.

To see this whole manual from the beginning in Info, you can run

```
$ info gnuastro
```

If you run Info with the particular program executable name, for example `astimgcrop` or `astnoisechisel`:

```
$ info astprogramname
```

you will be taken to the section titled "Invoking ProgramName" which explains the inputs and outputs along with the command line options for that program. Finally, if you run Info with the official program name, for example ImageCrop or NoiseChisel:

```
$ info ProgramName
```

you will be taken to the top section which introduces the program. Note that in all cases, Info is not case sensitive.

4.6.5 help-gnuastro mailing list

Gnuastro maintains the help-gnuastro mailing list for users to ask any questions related to Gnuastro. The experienced Gnuastro users and some of its developers are subscribed to this mailing list and your email will be sent to them immediately. However, when contacting this mailing list please have in mind that they are possibly very busy and might not be able to answer immediately.

To ask a question from this mailing list, send a mail to help-gnuastro@gnu.org. Anyone can view the mailing list archives at http://lists.gnu.org/archive/html/ help-gnuastro/. It is best that before sending a mail, you search the archives to see if anyone has asked a question similar to yours. If you want to make a suggestion or report a bug, please don't send a mail to this mailing list. We have other mailing lists and tools for those purposes, see Section 1.7 [Report a bug], page 7 or Section 1.8 [Suggest new feature], page 8.

4.7 Output headers

The output FITS files created by Gnuastro will have the following two keywords: DATE, CFITSIO, WCSLIB and GNUASTRO. The first specifies the time in UT of the file being created. The next three specify the versions of CFITSIO, WCSLIB and Gnuastro that was used to make the file. Note that WCSLIB has only recently added the version reporting capability. If you version of WCSLIB doesn't have this capability, it will not be reported. Some basic information about Gnuastro is also printed. The example below shows the last few keywords of one of the outputs of ImageCrop.

```
                 / ImageCrop (GNU Astronomy Utilities 0.1) 0.1:
DATE    = ' ... '                / file creation date ( ... )
CFITSIO = '3.37     '            / CFITSIO version.
WCSLIB  = '5.5      '            / WCSLIB version.
GNUASTRO= '0.1      '            / GNU Astronomy Utilities version.
COMMENT GNU Astronomy Utilities 0.1
COMMENT http://www.gnu.org/software/gnuastro/
END
```

5 Files

This chapter documents the programs in Gnuastro that are provided for getting information on the contents of a data file or converting a file format. Before working on a FITS file, it is commonly the case that you are not sure how many extensions it has within it and also what each extension is (image, table or blank). In other cases you want to use the data in a FITS file in other programs (for example in reports) that don't recognize the FITS format.

5.1 Header

The FITS standard requires each extension of a FITS file to have a header, giving basic information about what is in that extension. Each line in the header is for one keyword, specifying its name, value and a short comment string. Besides the basic information, the headers also contain vital information about the data, how they were processed, the instrument specifications that took the image and also the World Coordinate System that is used to translate pixel coordinates to sky or spectrum coordinates on the image or table.

5.1.1 Invoking Header

Header can print or manipulate the header information in an extension of an astronomical data file. The executable name is `astheader` with the following general template

```
$ astheader [OPTION...] ASTRdata
```

One line examples:

```
$ astheader image.fits
$ astheader --update=OLDKEY,153.034,"Old keyword comment"
$ astheader --remove=COMMENT --comment="Anything you like ;-)."
$ astheader --add=MYKEY1,20.00,"An example keyword" --add=MYKEY2,fd
```

If no keyword modification options are given, the full header of the given HDU will be printed on the command line. If any of the keywords are to be modified, the headers of the input file will be changed. If you want to keep the original FITS file, it is easiest to create a copy first and then run Header on that. In the FITS standard, keywords are always uppercase. So case does not matter in the input or output keyword names you specify.

Most of the options can accept multiple instances in one command. For example you can add multiple keywords to delete by calling delete multiple times, since repeated keywords are allowed, you can even delete the same keyword multiple times. The action of such options will start from the top most keyword.

> **FITS STANDARD KEYWORDS:** Some header keywords are necessary for later operations on a FITS file, for example BITPIX or NAXIS, see the FITS standard for their full list. If you modify (for example remove or rename) such keywords, the FITS file extension might not be usable any more. Also be careful for the world coordinate system keywords, if you modify or change their values, any future world coordinate system (like RA and Dec) measurements on the image will also change.

PRECEDENCE: The order of operations are as follows. Note that while the order within each class of actions is preserved, the order of individual actions is not. So irrespective of what order you called `--delete` and `--update`. First all the delete operations are going to take effect then the update operations.

1. `--delete`
2. `--rename`
3. `--update`
4. `--write`
5. `--asis`
6. `--history`
7. `--comment`
8. `--date`

All possible syntax errors will be reported before the keywords are actually written. FITS errors during any of these actions will be reported, but Header won't stop until all the operations are complete. If `quitonerror` is called, then Header will immediately stop upon the first error.

If only a certain set of header keywords are desired, it is easiest to pipe the output of Header to GNU Grep. Grep is a very powerful and advanced tool to search strings which is precisely made for such situations. For example if you only want to check the size of an image FITS HDU, you can run:

```
$ astheader input.fits | grep NAXIS
```

The options particular to Header can be seen below. See Section 4.1.4 [Common options], page 34 for a list of the options that are common to all Gnuastro programs, they are not repeated here.

`-a`

`--asis`   (=STR) Write the string within this argument exactly into the FITS file header with no modifications. If it does not conform to the FITS standards, then it might cause trouble, so please be very careful with this option. If you want to define the keyword from scratch, it is best to use the `--write` option (see below) and let CFITSIO worry about the standards. The best way to use this option is when you want to add a keyword from one FITS file to another unchanged and untouched. Below is an example of such a case that can be very useful sometimes:

```
$ key=$(astheader firstimage.fits | grep KEYWORD)
$ astheader --asis="$key" secondimage.fits
```

In particular note the double quotation signs (") around the reference to the `key` shell variable (`$key`), since FITS keywords usually have lots of space characters, if this variable is not enclosed within double quotation marks, the shell will only give the first word in the full keyword to this option, which will definitely be a non-standard FITS keyword and will make it hard to work on the file afterwords. See the "Quoting" section of the GNU Bash manual for more information if your keyword has the special characters `$`, `'`, or `\`.

`-d`

`--delete`   (=STR) Delete one instance of the desired keyword. Multiple instances of `--delete` can be given (possibly even for the same keyword). All keywords given will be removed from the headers in the opposite order (last given keyword will be deleted first). If the keyword doesn't exist, Header will give a warning and return with a non-zero value, but will not stop.

`-r`

`--rename`   (=STR) Rename a keyword to a new value. The old name and the new name should be separated by either a comma (`,`) or a space character. Note that if you use a space character, you have to put the value to this option within double quotation marks (`"`) so the space character is not interpreted as an option separator. Multiple instances of `--rename` can be given in one command. The keywords will be renamed in the specified order.

`-u`

`--update`   (=STR) Update a keyword, its value, its comments and its units all defined separately. If there are multiple instances of the keyword in the header, they will be changed from top to bottom (with multiple `--update` options).

The format of the values to this option can best be specified with an exmaple:

    --update=KEYWORD,value,"comments for this keyword",unit

The value can be any numerical or string value. Other than the `KEYWORD`, all the other values are optional. To leave a given token empty, follow the preceding comma (`,`) immediately with the next. If any space character is present around the commas, it will be considered part of the respective token. So if more than one token has space characters within it, the safest method to specify a value to this option is to put double quotation marks around each individual token that needs it. Note that without double quotation marks, space characters will be seen as option separators and can lead to undefined behavior.

`-w`

`--write`   (=STR) Write a keyword to the header. For the format of inputing the possible values, comments and units for the keyword, see the `--update` option above.

`-H`

`--history`

                (=STR) Add a `HISTORY` keyword to the header. The string given to this keyword will be separated into multiple keyword cards if it is longer than 70 characters. With each run only one value for the `--history` option will be read. If there are multiple, it is the last one.

`-c`

`--comment`

                (=STR) Add a `COMMENT` keyword to the header. Similar to the explanation for `--history` above.

`-t`

`--date`   Put the current date and time in the header. If the `DATE` keyword already exists in the header, it will be updated.

`-Q`

`--quitonerror`

> Quit if any of the operations above are not successful. By default if an error occurs, Header will warn the user of the faulty keyword and continue with the rest of actions.

5.2 ConvertType

The formats of astronomical data were defined mainly for archiving and processing. In other situations, the data might be useful in other formats. For example, when you are writing a paper or report or if you are making slides for a talk, you can't use a FITS image. Other image formats should be used. In other cases you might want your pixel values in a table format as plain text for input to other programs that don't recognize FITS, or to include as a table in your report. ConvertType is created for such situations. The various types will increase with future updates and based on need.

The conversion is not only one way (from FITS to other formats), but two ways (except the EPS and PDF formats). So you can convert a JPEG image or text file into a FITS image. Basically, other than EPS, you can use any of the recognized formats as different color channel inputs to get any of the recognized outputs. So before explaining the options and arguments, first a short description of the recognized files types will be given followed a short introduction to digital color.

5.2.1 Recognized file types

The various standards and the file name extensions recognized by ConvertType are listed below.

FITS or IMH

> Astronomical data are commonly stored in the FITS format (and in older data sets in IRAF `.imh` format), a list of file name suffixes which indicate that the file is in this format is given in Section 4.1.2 [Arguments], page 32.
>
> Each extension of a FITS image only has one value per pixel, so when used as input, each input FITS image contributes as one color channel. If you want multiple extensions in one FITS file for different color channels, you have to repeat the file name multiple times and use the `--hdu`, `--hdu2`, `--hdu3` or `--hdu4` options to specify the different extensions.

JPEG

> The JPEG standard was created by the Joint photographic experts group. It is currently one of the most commonly used image formats. Its major advantage is the compression algorithm that is defined by the standard. Like the FITS standard, this is a raster graphics format, which means that it is pixelated.
>
> A JPEG file can have 1 (for grayscale), 3 (for RGB) and 4 (for CMYK) color channels. If you only want to convert one JPEG image into other formats, there is no problem, however, if you want to use it in combination with other input files, make sure that the final number of color channels does not exceed four. If it does, then ConvertType will abort and notify you.
>
> The file name endings that are recognized as a JPEG file for input are: `.jpg`, `.JPG`, `.jpeg`, `.JPEG`, `.jpe`, `.jif`, `.jfif` and `.jfi`.

EPS   The Encapsulated PostScript (EPS) format is essentially a one page PostScript file which has a specified size. PostScript also includes non-image data, for example lines and texts. It is a fully functional programming language to describe a document. Therefore in ConvertType, EPS is only an output format and cannot be used as input. Contrary to the FITS or JPEG formats, PostScript is not a raster format, but is categorized as vector graphics.

The Portable Document Format (PDF) is currently the most common format for documents. Some believe that PDF has replaced PostScript and that PostScript is now obsolete. This view is wrong, a PostScript file is an actual plain text file that can be edited like any program source with any text editor. To be able to display its programmed content or print, it needs to pass through a processor or compiler. A PDF file can be thought of as the processed output of the compiler on an input PostScript file. PostScript, EPS and PDF were created and are registered by Adobe Systems.

With these features in mind, you can see that when you are compiling a document with TeX or LaTeX, using an EPS file is much more low level than a JPEG and thus you have much greater control and therefore quality. Since it also includes vector graphic lines we also use such lines to make a thin border around the image to make its appearance in the document much better. No matter the resolution of the display or printer, these lines will always be clear and not pixelated. In the future, addition of text might be included (for example labels or object IDs) on the EPS output. However, this can be done better with tools within TeX or LaTeX such as PGF/Tikz[1].

If the final input image (possibly after all operations on the flux explained below) is a binary image or only has two colors of black and white (in segmentation maps for example), then PostScript has another great advantage compared to other formats. It allows for 1 bit pixels (pixels with a value of 0 or 1), this can decrease the output file size by 8 times. So if a grayscale image is binary, ConvertType will exploit this property in the EPS and PDF (see below) outputs.

The standard formats for an EPS file are `.eps`, `.EPS`, `.epsf` and `.epsi`. The EPS outputs of ConvertType have the `.eps` suffix.

PDF   As explained above, a PDF document is a static document description format, viewing its result is therefore much faster and more efficient than PostScript. To create a PDF output, ConvertType will make a PostScript page description and convert that to PDF using GPL Ghostscript. The suffixes recognized for a PDF file are: `.pdf`, `.PDF`. If GPL Ghostscript cannot be run on the PostScript file, it will remain and a warning will be printed.

`blank`   This is not actually a file type! But can be used to fill one color channel with a blank value. If this argument is given for any color channel, that channel will not be used in the output.

Plain text   Plain text files have the advantage that they can be viewed with any text editor or on the command line. Most programs also support input as plain text files.

---

[1] http://sourceforge.net/projects/pgf/

In ConvertType, if the input arguments do not have any of the extensions listed above for other formats, the input is assumed to be a text file. Each plain text file is considered to contain one color channel. There is no standard output for plain text files.

If any of the extension above is mis-spelled, this will result in the output becoming a plain text file with that (short) name. If this happens, ConvertType will warn you and write the output as a plain text file. If you don't want that warning, set your plain text output file names longer than 5 characters. When converting an image to plain text, consider the fact that if the image is large the number of columns in each line will become very large, possibly making it very hard to open in some text editors.

5.2.2 Color

An image is a two dimensional array of 2 dimensional elements called pixels. If each pixel only has one value, the image is known as a grayscale image and no color is defined. The range of values in the image can be interpreted as shades of any color, it is customary to use shades of black or grayscale. However, to produce the color spectrum in the digital world, several primary colors must be mixed. Therefore in a color image, each pixel has several values depending on how many primary colors were choosen. For example on the digital monitor or color digital cameras, all colors are built by mixing the three colors of Red-Green-Blue (RGB) with various proportions. However, for printing on paper, standard printers use the Cyan-Magenta-Yellow-Key (CMYK, Key=black) color space. Therefore when printing an RGB image, usually a transformation of color spaces will be necessary.

In a colored digital camera, a color image is produced by dividing the pixel's area between three colors (filters). However in astronomy due to the intrinsic faintness of most of the targets, the collecting area of the pixel is very important for us. Hence the full area of the pixel is used and one value is stored for that pixel in the end. One color filter is used for the whole image. Thus a FITS image is inherently a grayscale image and no color can be defined for it.

One way to represent a grayscale image in different color spaces is to use the same proportions of the primary colors in each pixel. This is the common way most FITS image converters work: they fill all the channels with the same values. The downside is two fold:

- Three (for RGB) or four (for CMYK) values have to be stored for every pixel, this makes the output file very heavy (in terms of bytes).

- If printing, the printing errors of each color channel can make the printed image slightly more blurred than it actually is.

To solve both these problems, the best way is to save the FITS image into the black channel of the CMYK color space. In the RGB color space all three channels have to be used. The JPEG standard is the only common standard that accepts CMYK color space, that is why currently only the JPEG standard is included and not the PNG standard for example.

The JPEG and EPS standards set two sizes for the number of bits in each channel: 8-bit and 12-bit. The former is by far the most common and is what is used in ConvertType. Therefore, each channel should have values between 0 to $2^8 - 1 = 255$. From this we see how each pixel in a grayscale image is one byte (8 bits) long, in an RGB image, it is 3bytes long and in CMYK it is 4bytes long. But thanks to the JPEG compression algorithms, when all the pixels of one channel have the same value, that channel is compressed to one pixel. Therefore a Grayscale image and a CMYK image that has only the K-channel filled are approximately the same file size.

5.2.3 Invoking ConvertType

ConvertType will convert any recognized input file type to any specified output type. The executable name is `astconvertt` with the following general template

```
$ astconvertt [OPTION...] InputFile [InputFile2] ... [InputFile4]
```

One line examples:

```
$ astconvertt M31.fits --output=pdf
$ astconvertt galaxy.jpg -ogalaxy.fits
$ astconvertt f1.txt f2.txt f3.fits -o.jpg
$ astconvertt M31_r.fits M31_g.fits blank -oeps
```

The file type of the output will be specified with the (possibly complete) file name given to the `--output` option, which can either be given on the command line or in any of the configuration files (see Section 4.2 [Configuration files], page 37). Note that if the output suffix is not recognized, it will default to plain text format, see Section 5.2.1 [Recognized file types], page 51.

The order of multiple input files is important. After reading the input file(s) the number of color channels in all the inputs will be used to define which color space is being used for the outputs and how each color channel is interpreted. Note that one file might have more than one color channel (for example in the JPEG format). If there is one color channel the output is grayscale, if three input color channels are given they are respectively considered to be the red, green and blue color channels and if there are four color channels they are respectively considered to be cyan, magenta, yellow and black.

The value to `--output` (or `-o`) can be either a full file name or just the suffix of the desired output format. In the former case, that same name will be used for the output. In the latter case, the name of the output file will be set based on the automatic output guidelines, see Section 4.5 [Automatic output], page 43. Note that the suffix name can optionally start a . (dot), so for example `--output=.jpg` and `--output=jpg` are equivalent. Be careful that if you want your output in plain text, you have to give the full file name. So if `-otxt` or `--output=.txt` are given, the output file will be named `txt` or `.txt` (the latter will be a hidden file!).

Besides the common set of options explained in Section 4.1.4 [Common options], page 34, the options to ConvertType can be classified into input, output and flux related options. The majority of the options are to do with the flux range. Astronomical data usually have a very large dynamic range (difference between maximum and minimum value) and different subjects might be better demonstrated with a limited flux range.

Input:

`--hdu2`   If the second input file is a FITS file, the value to this option will be used to specify which HDU will be used. Note that for the first file, the (`--hdu` or `-h` in the common options is used)

`--hdu3`   The HDU of the third input FITS file.

`--hdu4`   The HDU of the fourth input FITS file.

Output:

`-w`
`--widthincm`

> (=FLT) The width of the output in centimeters. This is only relevant for those formats that accept such a width (not plain text for example). For most digital purposes, the number of pixels is far more important than the value to this parameter because you can adjust the absolute width (in inches or centimeters) in your document preparation program.

`-b`
`--borderwidth`

> (=INT) The width of the border to be put around the EPS and PDF outputs in units of PostScript points. There are 72 or 28.35 PostScript points in an inch or centimeter respectively. In other words, there are roughly 3 PostScript points in every millimeter. If you are planning on adding a border, its significance is highly correlated with the value you give to the `--widthincm` parameter.

> Unfortunately in the document structuring convention of the PostScript language, the "bounding box" has to be in units of PostScript points with no fractions allowed. So the border values only have to be specified in integers. To have a final border that is thinner than one PostScript point in your document, you can ask for a larger width in ConvertType and then scale down the output EPS or PDF file in your document preparation program. For example by setting `width` in your `includegraphics` command in TeX or LaTeX. Since it is vector graphics, the changes of size have no effect on the quality of your output quality (pixels don't get different values).

`-x`
`--hex`   Use Hexadecimal encoding in creating EPS output. By default the ASCII85 encoding is used which provides a much better compression rate. When converted to PDF (or included in TeX or LaTeX which is finally saved as a PDF file), an efficient binary encoding is used which is far more efficient than both of them. The choice of EPS encoding will thus have no effect on the final PDF.

> So if you want to transfer your EPS files (for example if you want to submit your paper to arXiv or journals in PostScript), their storage might become important if you have large images or lots of small ones. By default ASCII85 encoding is used which offers a much better compression rate (nearly 40 percent) compared to Hexadecimal encoding.

`-u`
`--quality`

> (=INT) The quality (compression) of the output JPEG file with values from 0 to 100 (inclusive). For other formats the value to this option is ignored. Note that only in grayscale (when one input color channel is given) will this actually be the exact quality (each pixel will correspond to one input value). If it is in color mode, some degradation will occur. While the JPEG standard does support loss-less graphics, it is not commonly supported.

Flux range:

-c

--change   (=STR) Change pixel values with the following format "from1:to1, from2:to2,...". This option is very useful in displaying labeled pixels (not actual data images which have noise) like segmentation maps. In labeled images, usually a group of pixels have a fixed integer value. With this option, you can manipulate the labels before the image is displayed to get a better output for print or to emphasize on a particular set of labels and ignore the rest. The labels in the images will be changed in the same order given. By default first the pixel values will be converted then the pixel values will be truncated (see --fluxlow and --fluxhigh).

You can use any number for the values irrespective of your final output, your given values are stored and used in the double precision floating point format. So for example if your input image has labels from 1 to 20000 and you only want to display those with labels 957 and 11342 then you can run ConvertType with these options:

```
$ astconvertt --change=957:50000,11342:50001 --fluxlow=5e4 \
    --fluxhigh=1e5 segmentationmap.fits --output=jpg
```

While the output JPEG format is only 8 bit, this operation is done in an intermediate step which is stored in double precision floating point. The pixel values are converted to 8-bit after all operations on the input fluxes have been complete. By placing the value in double quotes you can use as many spaces as you like for better readability.

-C

--changeaftertrunc

Change pixel values (with --change) after truncation of the flux values, by default it is the opposite.

-L

--fluxlow

(=FLT) The minimum flux (pixel value) to display in the output image, any pixel value below this value will be set to this value in the output. If the value to this option is the same as --fluxmax, then no flux truncation will be applied. Note that when multiple channels are given, this value is used for all the color channels.

-H

--fluxhigh

(=FLT) The maximum flux (pixel value) to display in the output image, see --fluxlow.

-m

--maxbyte

(=INT) This is only used for the JPEG and EPS output formats which have an 8-bit space for each channel of each pixel. The maximum value in each pixel can therefore be $2^8 - 1 = 255$. With this option you can change (decrease) the maximum value. By doing so you will decrease the dynamic range. It can be useful if you plan to use those values for other purposes.

`-i`
`--flminbyte`

> (=INT) If the lowest pixel value in the input channels is larger than the value to `--fluxlow`, then that input value will be redundant. In some situations it might be necessary to set the minimum byte value (0) to correspond to that flux even if the data do not reach that value. With this option you can do this. Note that if the minimum pixel value is smaller than `--fluxlow`, then this option is redundant.

`-a`
`--fhmaxbyte`

> (=INT) See `--flminbyte`.

`-l`
`--log`   Display the logarithm of the input data. This is done after the conversion and flux truncation steps, see above.

`-n`
`--noinvert`

> For 8-bit output types (JPEG and EPS for example) the final value that is stored is inverted so white becomes black and vice versa. The reason for this is that astronomical images usually have a very large area of blank sky in them. The result will be that a large are of the image will be black. Therefore, by default the 8-bit values are inverted so the images blend in better with the text in a document.

> Note that this behaviour is ideal for grayscale images, if you want a color image, the colors are going to be mixed up. For color images it is best to call this option so the image is not inverted.

6 Image manipulation

Images are one of the major formats of data that is used in astronomy. The functions in this chapter explain the GNU Astronomy Utilities which are provided for their manipulaton. For example cropping out a part of a larger image or convolving the image with a given kernel or applying a transformation to it.

6.1 ImageCrop

Astronomical images are often very large, filled with thousands of galaxies. It often happens that you only want a section of the image, or you have a catalog of sources and you want to visually analyze them in small postage stamps. ImageCrop is made to do all these things. When more than one crop is required, ImageCrop will divide the crops between multiple threads to significantly reduce the run time.

Astronomical surveys are usually extremely large. So large in fact, that the whole survey will not fit into a reasonably sized file. Because of this surveys usually cut the final image into separate tiles and store each tile in a file. For example the COSMOS survey's Hubble space telescope, ACS F814W image consists of 81 separate FITS images, with each one having a volume of 1.7 Giga bytes.

Even though the tile sizes are chosen to be large enough that too many galaxies don't fall on the edges of the tiles, inevitably some do and if you simply crop the image of the galaxy from that one tile, you will miss a large area of the surrounding sky (which is essential in estimating the noise). Therefore in its WCS mode, ImageCrop will stitch parts of the tiles that are relevant for a target (with the given width) from all the input images that cover that region into the output. Of course, the tiles have to be present in the list of input files.

ImageCrop also has facilities to crop arbitrary polygons from a set of tiles by stitching the relevant parts of different tiles within the polygon, see `--polygon` in Section 6.1.4 [Invoking ImageCrop], page 61. Alternatively, it can crop out rectangular regions from one image which can come in handy when removing the bias pixels in raw image processing, see Section 6.1.2 [Crop section syntax], page 60.

6.1.1 ImageCrop modes

In order to be as comprehensive as possible, ImageCrop has two major modes of operation listed below.

Image   The image mode uses the pixel coordinates. Depending on your command line options, this mode consists of three sub-modes. In image mode, only one image may be input.

- Catalog (multiple crops). Coordinates are read from a text file. The `--xcol` and `--ycol` columns in the catalog are interpreted as the center of a square crop box whose width is specified with the `--iwidth` option in pixels. Since the given pixel has to be on the center, the width has to be an odd number, so if you give an even number for the width, it will be added by one. If a catalog file name is provided (with `--imagemode` activated of course) this mode will be used.

- Center (one crop). The box center is given on the command line with the `--xc` and `--yc` parameters. The image width is similar to above.

- Section (one crop). You can specify the section of pixels along each axis in the image which you want to be cropped with the `--section` option. See Section 6.1.2 [Crop section syntax], page 60 for a full explanation on the syntax of specifying the desired region.

The latter two cases will only have one crop box. In both cases, ImageCrop will go into the image mode, irrespective of calling `--wcsmode` or the default mode. In the first two cases, since you specify a central pixel, the crop box will be a square with an odd number of pixels on the side, so your desired pixel sits right in the center, see Section 6.1.3 [Blank pixels], page 60 on how to disable this for cases when the box exceeds the image size.

WCS   The Right ascension (RA) and Declination (Dec) of the objects in a catalog is used to define the central position of each postage stamp. In this mode, the width (`--wwidth`) is read in units of arc seconds and multiple images (tiles in a survey) can be input. If the objects are closer to the edge of the image than half the required width, other tiles (if they are present in the input files) are used to fill the empty space. The square output cropped box will have an odd number of pixels on the side.

In this mode, the input images do not necessarily have to be the same size, each individual tile can even be smaller than the final crop. In any case, any part of any of the input images which overlaps with the desired region will be used in the crop. Note that if there is an over lap, the pixels from the last input image read are going to be used. The input images all just have to be aligned with the celestial coordinates, see the caution note below.

Similar to the image mode, there are two sub-modes:

- Catalog (multiple crops). Similar to catalog mode in image mode. The RA and Dec column should be specified in the catalog (`--racol` and `--deccol`).
- Center (one crop). You can specify the center of only one crop box (no matter how many input images there are) with the options `--ra` and `--dec`. If it exists in the input images, it will be cropped similar to the catalog mode. If automatic output is triggered (you don't specify a file name for `--output`) and several of the input images are used to stitch and crop the region around the central point, the name of the first input will be used in automatic output, see Section 4.5 [Automatic output], page 43.

> **CAUTION:** In WCS mode, the image has to be aligned with the celestial coordinates, such that the first FITS axis is parallel (opposite direction) to the Right Ascension (RA) while the second FITS axis is parallel to the declination. If these conditions aren't met for an image, ImageCrop will warn you and abort. You have to use other tools to transform the image to the correct directions.

In short, if you don't specify a catalog, you have to specify box coordinates manually on the command line. When you do specify a catalog, ImageCrop has to be in one of the two major modes (`--imgmode` or `--wcsmode`). Note that the single crop box parameters specified in the sub-modes will not be written to or read from the configuration file, they have to be specified on each execution.

6.1.2 Crop section syntax

When in image mode, one of the methods to crop only one rectangular section from the input image is to use the `--section` option. ImageCrop has a powerful syntax to read the box parameters from a string of characters. If you leave certain parts of the string to be empty, ImageCrop can fill them for you based on the input image sizes.

To define a box, you need the coordinates of two points: the first pixel in the box at (`X1`, `Y1`) and the pixel which is immediately outside of the box (`X2`, `Y2`), four coordinates in total. The four coordinates can be specified with one string in this format: `X1:X2,Y1:Y2`. It is given to the `--section` option. Therefore, the pixels along the first axis that are $\geq$`X1` and `<X2` will be included in the cropped image. The same goes for the second axis. Note that each different term will be read as an integer, not a float (there are no sub-pixels in ImageCrop, you can use ImageWarp to shift the matrix with any subpixel distance, then crop the warped image, see Section 6.3 [ImageWarp], page 86). Also, following the FITS standard, pixel indexes along each axis start from unity(1) not zero(0).

You can omit any of the values and they will be filled automatically. The left hand side of the colon (`:`) will be filled with 1, and the right side with the image size. So, `2:,:` will include the full range of pixels along the second axis and only those with a first axis index larger than 2 in the first axis. If the colon is omitted for a dimension, then the full range is automatically used. So the same string is also equal to `2:,` or `2:` or even `2`. If you want such a case for the second axis, you should set it to: `,2`.

If you specify a negative value, it will be seen as before the indexes of the image which are outside the image along the bottom or left sides when viewed in SAO ds9. In case you want to count from the top or right sides of the image, you can use an asterisk (`*`). When confronted with a `*`, ImageCrop will replace it with the maximum length of the image in that dimension. So `*-10:*+10,*-20:*+20` will mean that the crop box will be 20×40 pixels in size and only include the top corner of the input image with 3/4 of the image being covered by blank pixels, see Section 6.1.3 [Blank pixels], page 60.

If you feel more comfortable with space characters between the values, you can use as many space characters as you wish, just be careful to put your value in double quotes, for example `--section="5:200, 123:854"`. If you forget, anything after the first space will not be seen by `--section`, because the unquoted space character is one of the characters that separates options on the command line.

6.1.3 Blank pixels

The cropped box can potentially include pixels that are beyond the image range. For example when a target in the input catalog was very near the edge of the input image. The parts of the cropped image that were not in the input image will be filled with the following two values depending on the data type of the image. In both cases, SAO ds9 will not color code those pixels.

- If the data type of the image is a floating point type (float or double), IEEE NaN (Not a number) will be used.
- For integer types, pixels out of the image will be filled with the value of the `BLANK` keyword in the cropped image header. The value assigned to it is the lowest value

possible for that type, so you will probably never need it any way. Only for the unsigned character type (`BITPIX=8` in the FITS header), the maximum value is used because it is unsigned, the smallest value is zero which is often meaningful.

You can ask for such blank regions to not be included in the output crop image using the `--noblank` option. In such cases, there is no guarantee that the image size of your outputs are what you asked for.

In some survey images, unfortunately they do not use the `BLANK` FITS keyword. Instead they just give all pixels outside of the survey area a value of zero. So by default, when dealing with float or double image types, any values that are 0.0 are also regarded as blank regions. This can be turned off with the `--zeroisnotblank` option.

6.1.4 Invoking ImageCrop

ImageCrop will crop a region from an image. If in WCS mode, it will also stitch parts from separate images in the input files. The executable name is `astimgcrop` with the following general template

```
$ astimgcrop [OPTION...] [ASCIIcatalog] ASTRdata ...
```

One line examples:

```
$ astimgcrop -I catalog.txt image.fits
$ astimgcrop -W catalog.txt /mnt/data/COSMOS/*_drz.fits
$ astimgcrop --section=10:*-10,10:*-10 --hdu=2 image.fits
$ astimgcrop --ra=189.16704 --dec=62.218203 goodsnorth.fits
$ astimgcrop --xc=568.342 --yc=2091.719 --iwidth=200 image.fits
```

ImageCrop has one mandatory argument which is the input image name(s), shown above with `ASTRdata ...`. You can use shell expansions, for example * for this if you have lots of images in WCS mode. If the crop box centers are in a catalog, you also have to provide the catalog name as an argument. Alternatively, you have to provide the crop box parameters with command line options.

When in catalog mode, ImageCrop will run using any number of threads that you have specified with the `--numthreads` option, see Section 4.1.4 [Common options], page 34. Note that when multiple threads are being used, in verbose mode, the outputs will not be in order. This is because the threads are asynchronous and thus not started in order. When the box coordinates are given on the command line, no threads will be created.

6.1.4.1 ImageCrop options

The options can be classified into the following contexts: Input, Output and operating mode options. Options that are common to all Gnuastro program are listed in Section 4.1.4 [Common options], page 34 and will not be repeated here.

> **NOTE:** The coordinates are in the FITS format. So the first axis is the horizontal axis when viewed in SAO ds9 and the second axis is the vertical. Also in the FITS standard, counting begins from 1 (one) not 0 (zero).

Input image parameters:

`--hstartwcs`

> (=INT) Specify the first keyword card (line number) to start finding the input image world coordinate system information. Distortions were only recently included in WCSLIB (from version 5). Therefore until now, different telescope would apply their own specific set of WCS keywords and put them into the image header along with those that WCSLIB does recognize. So now that WCSLIB recognizes most of the standard distortion parameters, they will get confused with the old ones and give completely wrong results. For example in the CANDELS-GOODS South images[1].
>
> The two `--hstartwcs` and `--hendwcs` are thus provided so when using older datasets, you can specify what region in the FITS headers you want to use to read the WCS keywords. Note that this is only relevant for reading the WCS information, basic data information like the image size are read separately. These two options will only be considered when the value to `--hendwcs` is larger than that of `--hstartwcs`. So if they are equal or `--hstartwcs` is larger than `--hendwcs`, then all the input keywords will be parsed to get the WCS information of the image.

`--hendwcs`

> (=INT) Specify the last keyword card to read for specifying the image world coordinate system on the input images. See `--hstartwcs`

Crop box parameters:

`-x`

`--xc`
> (=FLT) The first FITS axis value of central position of the crop box in single image mode.

`-y`

`--yc`
> (=FLT) The second FITS axis value of the central position of the crop box in single image mode.

`-s`

`--section`
> (=STR) Section of the input image which you want to be cropped. See Section 6.1.2 [Crop section syntax], page 60 for a complete explanation on the syntax required for this input.

`-l`

`--polygon`
> (=STR) String of crop polygon vertices. Note that currently only convex polygons should be used. In the future we will make it work for all kinds of polygons. Convex polygons are polygons that do not have an internal angle more than 180 degrees. This option can be used both in the image and WCS modes. The rectangular region that completely encompasses the polygon will be kept and all the pixels that are outside of it will be removed.
>
> The syntax for the polygon vertices is similar to and simpler than that for `--section`. In short, the dimentions of each coordinate are separated by a

---

[1] https://archive.stsci.edu/pub/hlsp/candels/goods-s/gs-tot/v1.0/

comma (,) and each vertice is separated by a colon (:). You can define as many vertices as you like. If you would like to use space characters between the dimentions and vertices to make them more human-readible, then you have to put the value to this option in double quotation marks.

For example let's assume you want to work on the deepest part of the WFC3/IR images of Hubble Space Telescope eXtreme Deep Field (HST-XDF). According to the webpage[2] the deepest part is contained within the coordinates:

```
[ (53.187414,-27.779152), (53.159507,-27.759633),
    (53.134517,-27.787144), (53.161906,-27.807208) ]
```

They have provided mask images with only these pixels in the WFC3/IR images, but what if you also need to work on the same region in the full resolution ACS images? Also what if you want to use the CANDELS data for the shallow region? Running ImageCrop with --polygon will easily pull out this region of the image for you irrespective of the resolution (if you have set the operating mode to WCS mode in your nearest configuration file, there is no need for --wcsmode, you may also provide many FITS image or tiles and ImageCrop will stitch them all):

```
$ astimgcrop --wcsmode desired-filter-image(s).fits         \
    --polygon="53.187414,-27.779152 : 53.159507,-27.759633 : \
        53.134517,-27.787144 : 53.161906,-27.807208"
```

--outpolygon

Keep all the regions outside the polygon and mask the inner ones with blank pixels (see Section 6.1.3 [Blank pixels], page 60). This is practically the inverse of the default mode of treating polygons. Note that this option only works when you have only provided one input image. If multiple images are given (in WCS mode), then the full area covered by all the images has to be shown and the polygon excluded. This can lead to a very large area if large surveys like COSMOS are used. So ImageCrop will abort and notify you. In such cases, it is best to crop out the larger region you want, then mask the smaller region with this option.

-r
--ra   (=FLT) The first FITS axis value of central position of the crop box in single image mode.

-d
--dec   (=FLT) The second FITS axis value of the central position of the crop box in single image mode.

-i
--xcol   (=INT) Column number of the first FITS axis position of the box center, starting from zero. In SAO ds9, the first FITS axis is the horizontal axis.

-j
--ycol   (=INT) Column number of the second FITS axis position of the box center, starting from zero. In SAO ds9, the second FITS axis is the vertical axis.

---

[2] https://archive.stsci.edu/prepds/xdf/

```
-a
--iwidth
```
(=INT) Width the square box to crop in image mode in units of pixels. In order for the chosen central pixel to be in the center of the cropped image, the final width has to be an odd number, therefore if the value to this option is an even number, the final crop width will be one pixel larger in each dimention. If you want an even sided crop box, use the **--section** option to specify the boudaries of the box, see Section 6.1.2 [Crop section syntax], page 60.

```
-f
--racol
```
(=INT) Column number of Right Ascension (RA) in the input catalog, starting from zero.

```
-g
--deccol
```
(=INT) Column number of declination in the input catalog, starting from zero.

```
-w
--wwidth
```
(=FLT) The width of the crop box in WCS mode in units of arc-seconds.

Output options:

```
-c
--checkcenter
```
(=INT) Box size of region in the center of the image to check in units of pixels. This is only used in WCS mode. Because surveys don't often have a clean square or rectangle shape, some of the pixels on the sides of the surveys don't have any data and are commonly filled with zero valued pixels.

If the RA and Dec of any of the targets specified in the catalog fall in such regions, that cropped image will be useless! Therefore with this option, you can specify a width of a small box (3 pixels is often good enough) around the central pixel of the cropped image. If all the pixels in this small box have the value of zero, no cropped image will be created and this object will be flagged in the final log file.

```
-p
--suffix
```
(=STR) The suffix (or post-fix) of the output files for when you want all the cropped images to have a special ending. One case where this might be helpful is when besides the science images, you want the weight images (or exposure maps, which are also distributed with survey images) of the cropped regions too. So in one run, you can set the input images to the science images and **--suffix=_s.fits**. In the next run you can set the weight images as input and **--suffix=_w.fits**.

```
-b
--noblank
```
Pixels outside of the input image that are in the crop box will not be used. By default they are filled with blank values (depending on type), see Section 6.1.3 [Blank pixels], page 60.

```
-z
--zeroisnotblank
```
In float or double images, it is common to give the value of zero to blank pixels. If the input image type is one of these two types, such pixels will also

be considered as blank. You can disable this behavior with this option, see Section 6.1.3 [Blank pixels], page 60.

Operating mode options:

`-I`

`--imgmode`

> Operate in Image mode as described above. This option is only useful when catalog is being provided. If coordinates are given on the command line, the mode is automatically set based on them.

`-W`

`--wcsmode`

> Operate in WCS mode. See explanations for `--imgmode`.

6.1.4.2 ImageCrop output

When a catalog is given, the value of `--output` (see Section 4.1.4 [Common options], page 34) will be seen as the directory to store the output cropped images. In such cases, the outputs will consist of two parts: a variable part (the row number of each target starting from 1) along with a fixed string which you can set with the `--suffix` option. Note that in catalog mode, only one image can be input.

When the crop box is specified on the command line, the value to `--output` will be used as a file name. If no output is specified or if it is a directory, the output file name will follow the automatic output names of Gnuastro, see Section 4.5 [Automatic output], page 43 for the input image.

The header of each output cropped image will contain the names of the input image(s) it was cut from. If a name is longer than the 70 character space that the FITS standard allows for header keyword values, the name will be cut into several keywords from the nearest slash (/). The keywords have the following format: `ICFn_m`. Where `n` is the number of the image used in this crop and `m` is the part of the name. Following the name is another keyword named `ICFnPIX` which shows the pixel range from that input image in the same syntax as Section 6.1.2 [Crop section syntax], page 60.

Once done, a log file will be created in the current directory named `astimgcrop.log`. This file will keep the names of all the outputs along with the number of images that were used in them and also whether the central pixels of the cropped image are full. There are also comments on the top explaining basic information about the run. If the log file cannot be created (for example you don't have write permission in the directory you are running ImageCrop in) it will not be created (unless `--individual` is called). You can see the same results in verbose mode on the command line in such cases.

6.2 Convolve

On an image, convolution can be thought of as a process to blur or remove the contrast in an image. If you are already familiar with the concept and just want to run Convolve, you can jump to Section 6.2.4 [Convolution kernel], page 83 and Section 6.2.5 [Invoking Convolve], page 84 and skip the lengthy introduction on the basic definitions and concepts of convolution.

There are generally two methods to convolve an image. The first and more intuitive one is in the "spatial domain" or using the actual image pixel values, see Section 6.2.1 [Spatial domain convolution], page 66. The second method is when we manipulate the "frequency domain", or work on the magnitudes of the different frequencies that constitute the image, see Section 6.2.2 [Frequency domain and Fourier operations], page 68. Understanding convolution in the spatial domain is more intuitive and thus recommended if you are just starting to learn about convolution. However, getting a good grasp of the frequency domain is a little more involved and needs some concentration and some mathematical proofs. However, its reward is a faster operation and more importantly a very fundamental understanding of this very important operation.

Convolution of an image will generally result in blurring the image because it mixes pixel values. In other words, if the image has sharp differences in neighboring pixel values[3], those sharp differences will become smoother. This has very good consequences in detection of signal in noise for example. In an actual observed image, the variation in neighboring pixel values due to noise can be very high. But after convolution, those variations will decrease and we have a better hope in detecting the possible underlying signal. Another case where convolution is extensively used is in mock images and modelling in general, convolution can be used to simulate the effect of the atmosphere or the optical system on the mock profiles that we create, see Section 8.1.1.2 [Point Spread function], page 130. Convolution is a very interesting and important topic in any form of signal analysis (including astronomical observations). So we have thoroughly[4] explained the concepts behind it in the following sub-sections.

6.2.1 Spatial domain convolution

The pixels in an input image represent different "spatial" positions, therefore when convolution is done only using the actual input pixel values, we name the process as being done in the "Spatial domain". In particular this is in contrast to the "frequency domain" that we will discuss later in Section 6.2.2 [Frequency domain and Fourier operations], page 68. In the spatial domain (and in realistic situations where the image and the convolution kernel don't extend to infinity), convolution is the process of changing the value of one pixel to the *weighted* average of all the pixels in its *neighborhood*.

The 'neighborhood' of each pixel (how many pixels in which direction) and the 'weight' function (how much each neighboring pixel should contribute depending on its position) are given through a second image which is known as a "kernel"[5].

6.2.1.1 Convolution process

In convolution, the kernel specifies the weight and positions of the neighbors of each pixel. To find the convolved value of a pixel, the central pixel of the kernel is placed on that pixel. The values of each overlapping pixel in the kernel and image are multiplied by each other and summed for all the kernel pixels. To have one pixel in the center, the sides of

---

[3] In astronomy, the only major time we confront such sharp borders in signal are cosmic rays. All other sources of signal in an image are already blurred by the atmosphere or the optics of the instrument.

[4] A mathematicial will certainly consider this explanation is incomplete and inaccurate. However this text is written for an understanding on the operations that are done on a real (not complex, discrete and noisy) astronomical image, not any general form of abstract function

[5] Also known as filter, here we will use 'kernel'.

the convolution kernel have to be an odd number. This process effectively mixes the pixel values of each pixel with its neighbors, resulting in a blurred image compared to the sharper input image.

Formally, convolution is one kind of linear 'spatial filtering' in image processing texts. If we assume that the kernel has $2a+1$ and $2b+1$ pixels on each side, the convolved value of a pixel placed at x and y ($C_{x,y}$) can be calculated from the neighboring pixel values in the input image (I) and the kernel (K) from

$$C_{x,y} = \sum_{s=-a}^{a} \sum_{t=-b}^{b} K_{s,t} \times I_{x+s,y+t}.$$

Any pixel coordinate that is outside of the image in the equation above will be considered to be zero. When the kernel is symmetric about its center the blurred image has the same orientation as the original image. However, if the kernel is not symmetric, the image will be affected in the opposite manner, this is a natural consequence of the definition of spatial filtering. In order to avoid this we can rotate the kernel about its center by 180 degrees so the convolved output can have the same original orentation. Technically speaking, only if the kernel is flipped the process is known *Convolution*. If it isn't it is known as *Correlation*.

To be a weighted average, the sum of the weights (the pixels in the kernel) have to be unity. This will have the consequence that the convolved image of an object and unconvolved object will have the same brightness (see Section 8.1.3 [Flux Brightness and magnitude], page 134), which is natural, because convolution should not eat up the object photons, it only disperses them.

6.2.1.2 Edges in the spatial domain

In purely 'linear' spatial filtering (convolution), there are problems on the edges of the input image. Here we will explain the problem in the spatial domain, see Section 6.2.2.10 [Edges in the frequency domain], page 82. The problem originates from the fact that on the edges, in practice[6], the sum of the weights we use on the actual image pixels is not unity. For example, as discussed above, a profile in the center of an image will have the same brightness before and after convolution. However, for a profile on the edge of the image, the brightness (sum of its pixel fluxes within the image, see Section 8.1.3 [Flux Brightness and magnitude], page 134) will not be equal, some of the flux is going to be 'eaten' by the edges.

If you ran `$ make check` on the source files of Gnuastro, you can see the this effect by comparing the `convolve_frequency.fits` with `convolve_spatial.fits` in the `./tests/` directory. In the spatial domain, by default, no assumption will be made about pixels outside of the image or any blank pixels in the image, see Section 6.1.3 [Blank pixels], page 60. The problem explained above will also occur on the sides of blank regions (which might be masked for example). The solution to this edge effect problem is only possible in the spatial domain. We have to discard the assumption that the sum of the kernel pixels is unity during the convolution process[7]. So taking W as the sum of the kernel pixels that

---

[6] Because we assumed the overlapping pixels outside the input image have a value of zero.

[7] ofcourse the sum of the kernel pixels still have to be unity.

used non-blank and in-image pixels, the equation in Section 6.2.1.1 [Convolution process], page 66 will become:

$$C_{x,y} = \frac{\sum_{s=-a}^{a} \sum_{t=-b}^{b} K_{s,t} \times I_{x+s,y+t}}{W}.$$

In this manner, objects which are near the sides of the image or blank pixels will also have the same brightness (within the image) before and after convolution. This correction is applied by default in Convolve when convolving in the spatial domain. To disable it, you can use the `--noedgecorrection` option. In the frequency domain, there is no way to avoid this loss of flux near the edges of the image, see Section 6.2.2.10 [Edges in the frequency domain], page 82.

Note that the edge effect discussed here is different from the one in Section 8.1.2 [If convolving afterwards], page 134. In making mock images we want to simulate a real observation. In a real observation the images of the galaxies on the sides of the CCD are first blurred by the atmosphere and instrument, then imaged. So light from the parts of a galaxy which are immediately outside the CCD will affect the parts of the galaxy which are covered by the CCD. Therefore in modeling the observation, we have to convolve an image that is larger than the input image by exactly half of the convolution kernel. We can hence conclude that this correction for the edges is only useful when working on actual observed images (where we don't have any more data on the edges) and not in modeling.

6.2.2 Frequency domain and Fourier operations

Getting a good grip on the frequency domain is usually not an easy job! So we have decided to give the issue a complete review here. Convolution in the frequency domain (see Section 6.2.2.6 [Convolution theorem], page 75) heavily relies on the concepts of Fourier transform (Section 6.2.2.4 [Fourier transform], page 73) and Fourier series (Section 6.2.2.3 [Fourier series], page 71) so we will be investigating these important operations first. It has become something of a cliché for people to say that the Fourier series "is a way to represent a (wave-like) function as the sum of simple sine waves" (from Wikipedia). However, sines themselves are abstract functions, so this statement really adds no extra layer of physical insight.

Before jumping head-first into the equations and proofs we will begin with a historical background to see how the importance of frequencies actually roots in our ancient desire to see everything in terms of circles. A short review of how the complex plane should be interpretted is then given. Having paved the way with these two basics, we define the Fourier series and subsequently the Fourier transform. Our final aim is to explain discrete Fourier transform, however some very important concepts need to be solidified first: The Dirac comb, convolution theorem and sampling theorem. So each of these topics are explained in their own separate sub-sub-section before going on to the discrete Fourier transform. Finally we revisit (after Section 6.2.1.2 [Edges in the spatial domain], page 67) the problem of convolution on the edges, but this time in the frequency domain. Understanding the sampling theorem and the discrete Fourier transform is very important in order to be able to pull out valuable science from the "discrete" image pixels. Therefore we have included the mathematical proofs and figures so you can have a clear understanding of these very important concepts.

6.2.2.1 Fourier series historical background

Ever since the ancient times, the circle has been (and still is) the simplest shape for abstract comprehention. All you need is a center point and a radius and you are done. All the points on a circle are at a fixed distance from the center. However, the moment you try to connect this elegantly simple and beautiful abstract construct (the circle) with the real world (for example compute its area or its circumference), things become really hard (ideally, impossible) because the irrational number π gets involved.

The key to understanding the Fourier series (thus the Fourier transform and finally the Discrete Fourier Transform) is our ancient desire to express everthing in terms of circles or the most exceptionally simple and elegant abstract human construct. Most people prefer to say the same thing in a more ahistorical manner: to break a function into sines and cosines. As the term "ancient" in the previous sentence implies, Jean-Baptiste Joseph Fourier (1768 – 1830 A.D.) was not the first person to do this. The main reason we know this process by his name today is that he came up with an ingenious method to find the necessary coefficients (radius of) and frequencies ("speed" of rotation on) the circles for any generic (integrable) function.

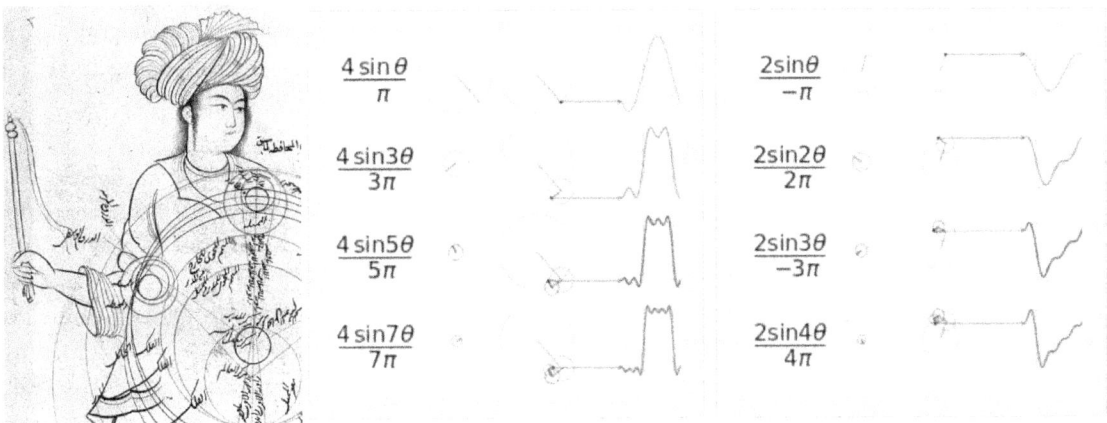

Figure 6.1: Epicycles and the Fourier series. Left: A demonstration of Mercury's epicycles relative to the "center of the world" by Qutb al-Din al-Shirazi (1236 – 1311 A.D.) retrieved from Wikipedia (`https://commons.wikimedia.org/wiki/File:Ghotb2.jpg`). Middle and Right: A snapshot from an animation Showing how adding more epicycles (or terms in the fourier series) will be able to approximate any function. Animations can be found at: (`https://commons.wikimedia.org/wiki/File:Fourier_series_square_wave_circles_animation.gif`) and (`https://commons.wikimedia.org/wiki/File:Fourier_series_sawtooth_wave_circles_animation.gif`).

Like most aspects of mathematics, this process of interpretting everything in terms of circles, began for astronomical purposes. When astronomers noticed that the orbit of Mars and other outer planets, did not appear to be a simple circle (as everything should have been in the heavens). At some point during their orbit, the revolution of these planets would become slower, stop, go back a little (in what is known as the retrograde motion) and then continue going forward again.

The correction proposed by Ptolemy (90 – 168 A.D.) was the most agreed upon. He put the planets on Epicycles or circles whose center itself rotates on a circle whose center is the earth. Eventually, as observations became more and more precise, it was necessary to add more and more epicycles in order to explain the complex motions of the planets[8]. Figure 6.1(Left) shows an example depiction of the epicycles of Mercury in the late 13th century.

Ofcourse we now know that if they had abdicated the Earth from its throne in the center of the heavens and allowed the Sun to take its place, everything would become much simpler and true. But there wasn't enough observational evidence for changing the "professional consensus" of the time to this radical view suggested by a small minority[9]. So the pre-Galilean astronomers chose to keep Earth in the center and find a correction to the models (while keeping the heavens a purely "circular" order).

The main reason we are giving this historical background which might appear off topic is to give historical evidence that while such "approximations" do work and are very useful for pragmatic reasons (like measuring the calendar from the movement of astronomical bodies). They offer no physical insight. The astronomers who were involved with the Ptolemic world view had to add a huge number of epicycles during the centuries after Ptolemy in order to explain more accurate observations. Finally the death knell of this world-view was Galileo's observations with his new instrument (the telescope). So the main physical insight which is what Astronomers and Physicists are interested in (as opposed to Mathematicians and Engineers who just like proving and optimizing or calculating!) comes from being creative and not limiting our selves to such approximations. Even when they work.

6.2.2.2 Circles and the complex plane

Before going onto the derivation, it is also useful to review how the complex numbers and their plane relate to the circles we talked about above. The two schematics in the middle and right of Figure 6.1 show how a 1D function of time can be made using the 2D real and imaginary surface. Seeing the animation in Wikipedia will really help in understanding this important concept. At each point in time, we take the vertical coordinate of the point and use it to find the value of the function at that point in time. Figure 6.2 shows this relation with the axises marked.

Leonhard Euler[10] (1707 – 1783 A.D.) showed that the complex exponential (e^{it} where t is real) is periodic and can be written as: $e^{it} = \cos t + i\sin t$. Therefore, $e^{it+2\pi} = e^{it}$. Since it is periodic (lets assume with a period of T), it is customary to write a complex exponential in the form of $e^{i\frac{2\pi n}{T}t}$ where n is an integer) instead of simply e^{it}. The advantage is of this notation is that the period (T) is clearly visible and the frequency ($\frac{2\pi n}{T}$) is defined through the integer n. In this notation, t is in units of "cycles". Later, Caspar Wessel (mathematician and cartographer 1745 – 1818 A.D.) showed how complex numbers can be

---

[8] See the Wikipedia page on "Deferent and epicycle" for a more complete historical review.

[9] Aristarchus of Samos (310 – 230 B.C.) appears to be one of the first peole to suggest the Sun being in the center of the universe. This approach to science (that the standard model is defined by concensus) and the fact that this consensus might be completely wrong still applies equally well to our models of particle physics and cosmology today.

[10] Other forms of this equation were known before Euler. For example in 1707 A.D. (the year of Euler's birth) Abraham de Moivre (1667 – 1754 A.D.) showed that $(\cos t + i\sin t)^n = \cos(nt) + i\sin(nt)$. In 1714 A.D., Roger Cotes (1682 – 1716 A.D. a colleague of Newton who proofread the second edition of Principia) showed that: $it = \ln(\cos t + i\sin t)$.

displayed as vectors on a plane and therefore how e^{it} can be interpretted as an angle on a circle.

As we see from the examples in Figure 6.1 and Figure 6.2, for each constituting frequency, we need a respective 'magnitude' or the radius of the circle in order to accurately approximate the desired 1D function. The concepts of "period" and "frequency" are relatively easy to grasp when using temporal units like time because this is how we define them in every-day life. However, in an image (astronomical data), we are dealing with spatial units like distance. Therefore, by one "period" we mean the *distance* at which the signal is identical and frequency is defined as the inverse of that spatial "period". The complex circle of Figure 6.2 resembles the Moon rotating about Earth which is rotating around the Sun; so the "Real (signal)" axis shows the Moon's position as seen by a distant observer positioned on $-\infty$ on the Imaginary (i) axis as time goes by. Therefore, because of the scalar (not having any direction or vector) nature of time, Figure 6.2 is easier to understand in units of time. When thinking about spatial units, mentally replace the "Time (sec)" axis with "Distance (meters)". Because length has direction and is a vector, making a connection between the rotation of the imaginary circle and the advance along the "Distance (meters)" axis is not as simple as temporal units like time.

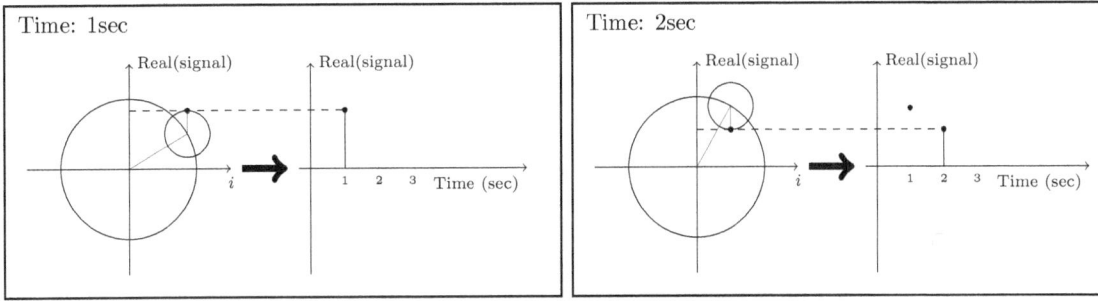

Figure 6.2: Relation between the real (signal), imaginary ($i \equiv \sqrt{-1}$) and time axises at two snapshots of time.

6.2.2.3 Fourier series

In astronomical images, our variable (brightness, or number of photo-electrons, or signal to be more generic) is recorded over the 2D spatial surface of a camera pixel. However to make things easier to understand, here we will assume that the signal is recorded in 1D (assume one row of the 2D image pixels). Also for this section and the next (Section 6.2.2.4 [Fourier transform], page 73) we will be talking about the signal before it is digitized or pixelated. Let's assume that we have the continuous function $f(l)$ which is integrable in the interval $[l_0, l_0 + L]$ (always true in practical cases like images). Take l_0 as the position of the first pixel in the assumed row of the image and L as the width of the image along that row. The units of l_0 and L can be in any spatial units (for example meters) or an angular unit (like radians) multiplied by a fixed distance which is more common.

To approximate $f(l)$ over this interval, we need to find a set of frequencies and their corresponding 'magnitude's (see Section 6.2.2.2 [Circles and the complex plane], page 70). Therefore our aim is to show $f(l)$ as the following sum of periodic functions:

$$f(l) = \sum_{n=-\infty}^{\infty} c_n e^{i\frac{2\pi n}{L}l}$$

Note that the different frequencies ($2\pi n/L$, in units of cycles per meters for example) are not arbitrary. They are all integer multiples of the fundamental frequency of $\omega_0 = 2\pi/L$. Recall that L was the length of the signal we want to model. Therefore, we see that the smallest possible frequency (or the frequency resolution) in the end, depends on the length we observed the signal or L. In the case of each dimension on an image, this is the size of the image in the respective dimension. The frequencies have been defined in this "harmonic" fashion to insure that the final sum is periodic outside of the $[l_0, l_0 + L]$ interval too. At this point, you might be thinking that the sky is not periodic with the same period as my camera's view angle. You are absolutely right! The important thing is that since your camera's observed region is the only region we are "observing" and will be using, the rest of the sky is irrelevant; so we can safely assume the is periodic outside of it. However, this working assumption will haunt us later in Section 6.2.2.10 [Edges in the frequency domain], page 82.

The frequencies are thus determined by definition. So all we need to do is to find the coefficients (c_n), or magnitudes, or radii of the circles for each frequency which is identified with the integer n. Fourier's approach was to multiply both sides with a fixed term:

$$f(l)e^{-i\frac{2\pi m}{L}l} = \sum_{n=-\infty}^{\infty} c_n e^{i\frac{2\pi(n-m)}{L}l}$$

where $m > 0$[11]. We can then integrate both sides over the observation period:

$$\int_{l_0}^{l_0+L} f(l)e^{-i\frac{2\pi m}{L}l}dl = \int_{l_0}^{l_0+L} \sum_{n=-\infty}^{\infty} c_n e^{i\frac{2\pi(n-m)}{L}l}dl = \sum_{n=-\infty}^{\infty} c_n \int_{l_0}^{l_0+L} e^{i\frac{2\pi(n-m)}{L}l}dl$$

Both n and m are positive integers. Also, we know that a complex exponential is periodic so after one period (L) it comes back to its starting point. Therefore $\int_{l_0}^{l_0+L} e^{2\pi k/L}dl = 0$ for any $k > 0$. However, when $k = 0$, this integral becomes: $\int_{l_0}^{l_0+T} e^0 dt = \int_{l_0}^{l_0+T} dt = T$. Hence since the integral will be zero for all $n \neq m$, we get:

$$\sum_{n=-\infty}^{\infty} c_n \int_{l_0}^{l_0+T} e^{i\frac{2\pi(n-m)}{L}l}dl = Lc_m$$

The origin of the axis is fundamentally an arbitrary position. So let's set it to the start of the image such that $l_0 = 0$. So we can find the "magnitude" of the frequency $2\pi m/L$ within $f(l)$ through the relation:

$$c_m = \frac{1}{L}\int_0^L f(l)e^{-i\frac{2\pi m}{L}l}dl$$

---

[11] We could have assumed $m < 0$ and set the exponential to positive, but this is more clear.

6.2.2.4 Fourier transform

In Section 6.2.2.3 [Fourier series], page 71, we had to assume that the function is periodic outside of the desired interval with a period of L. Therefore, assuming that $L \to \infty$ will allow us to work with any function. However, with this approximation, the fundamental frequency (ω_0) or the frequency resolution that we discussed in Section 6.2.2.3 [Fourier series], page 71 will tend to zero: $\omega_0 \to 0$. In the equation to find c_m, every m represented a frequency (multiple of ω_0) and the integration on l removes the dependance of the right side of the equation on l, making it only a function of m or frequency. Let's define the following two variables:

$$\omega \equiv m\omega_0 = \frac{2\pi m}{L}$$

$$F(\omega) \equiv Lc_m$$

The equation to find the coefficients of each frequency in Section 6.2.2.3 [Fourier series], page 71 thus becomes:

$$F(\omega) = \int_{-\infty}^{\infty} f(l)e^{-i\omega l}dl.$$

The function $F(\omega)$ is thus the *Fourier transform* of $f(l)$ in the frequency domain. So through this transformation, we can find (analyze) the magnitudes of the constituting frequencies or the value in the frequency space[12] of our spatial input function. The great thing is that we can also do the reverse and later synthesize the input function from its Fourier transform. Let's do it: with the approximations above, multiply the right side of the definition of the Fourier Series (Section 6.2.2.3 [Fourier series], page 71) with $1 = L/L = (\omega_0 L)/(2\pi)$:

$$f(l) = \frac{1}{2\pi} \sum_{n=-\infty}^{\infty} Lc_n e^{\frac{2\pi in}{L}l}\omega_0 = \frac{1}{2\pi} \sum_{n=-\infty}^{\infty} F(\omega)e^{i\omega l}\Delta\omega$$

To find the right most side of this equation, we renamed ω_0 as $\Delta\omega$ because it was our resolution, $2\pi n/L$ was written as ω and finally, Lc_n was written as $F(\omega)$ as we defined above. Now, as $L \to \infty$, $\Delta\omega \to 0$ so we can write:

$$f(l) = \frac{1}{2\pi} \int_{-\infty}^{\infty} F(\omega)e^{i\omega l}d\omega$$

Together, these two equations provide us with a very powerful set of tools that we can use to process (analyze) and recreate (synthesize) the input signal. Through the first equation, we can break up our input function into its constituent frequencies and analyze it, hence it

---

[12] As we discussed before, this 'magnitude' can be interpreted as the radius of the circle rotating at this frequency in the epicyclic interpretation of the Fourier series, see Figure 6.1 and Figure 6.2.

is also known as *analysis*. Using the second equation, we can synthesize or make the input function from the known frequencies and their magnitudes. Thus it is known as *synthesis*. Here, we symbolize the Fourier transform (analysis) and its inverse (synthesis) of a function $f(l)$ and its Fourier Transform $F(\omega)$ as $\mathcal{F}[f]$ and $\mathcal{F}^{-1}[F]$.

6.2.2.5 Dirac delta and comb

The Dirac δ (delta) function (also known as an impulse) is the way that we convert a continuous function into a discrete one. It is defined to satisfy the following integral:

$$\int_{-\infty}^{\infty} \delta(l)dl = 1$$

When integrated with another function, it gives that function's value at $l = 0$:

$$\int_{-\infty}^{\infty} f(l)\delta(l)dt = f(0)$$

An impulse positioned at another point (say l_0) is written as $\delta(l - l_0)$:

$$\int_{-\infty}^{\infty} f(l)\delta(l - l_0)dt = f(l_0)$$

The Dirac δ function also operates similarly if we use summations instead of integrals. The Fourier transform of the delta function is:

$$\mathcal{F}[\delta(l)] = \int_{-\infty}^{\infty} \delta(l)e^{-i\omega l}dl = e^{-i\omega 0} = 1$$

$$\mathcal{F}[\delta(l - l_0)] = \int_{-\infty}^{\infty} \delta(l - l_0)e^{-i\omega l}dl = e^{-i\omega l_0}$$

From the definition of the Dirac δ we can also define a Dirac comb (III_P) or an impulse train with infinite impules separated by P:

$$\mathrm{III}_P(l) \equiv \sum_{k=-\infty}^{\infty} \delta(l - kP)$$

P is chosen to represent "pixel width" later in Section 6.2.2.7 [Sampling theorem], page 77. Therefore the Dirac comb is periodic with a period of P. We have intentionally used a different name for the period of the Dirac comb compared to the input signal's length of observation that we showed with L in Section 6.2.2.3 [Fourier series], page 71. This difference is highlighted here to avoid confusion later when these two periods are needed together in Section 6.2.2.8 [Discrete Fourier transform], page 79. The Fourier transform of the Dirac comb will be necessary in Section 6.2.2.7 [Sampling theorem], page 77, so let's derive it. By its definition, it is periodic, with a period of P, so the Fourier coefficients of its Fourier Series (Section 6.2.2.3 [Fourier series], page 71) can be calculated within one period:

$$\text{III}_P = \sum_{n=-\infty}^{\infty} c_n e^{i\frac{2\pi n}{P}l}$$

We can now find the c_n from Section 6.2.2.3 [Fourier series], page 71:

$$c_n = \frac{1}{P}\int_{-P/2}^{P/2} \delta(l)e^{-i\frac{2\pi n}{P}l} = \frac{1}{P} \qquad \rightarrow \qquad \text{III}_P = \frac{1}{P}\sum_{n=-\infty}^{\infty} e^{i\frac{2\pi n}{P}l}$$

So we can write the Fourier transform of the Dirac comb as:

$$\mathcal{F}[\text{III}_P] = \int_{-\infty}^{\infty}\text{III}_P e^{-i\omega l}dl = \frac{1}{P}\sum_{n=-\infty}^{\infty}\int_{-\infty}^{\infty} e^{-i(\omega-\frac{2\pi n}{P})l}dl = \frac{1}{P}\sum_{n=-\infty}^{\infty}\delta\left(\omega - \frac{2\pi n}{P}\right)$$

In the last step, we used the fact that the complex exponential is a periodic function, that n is an integer and that as we defined in Section 6.2.2.4 [Fourier transform], page 73, $\omega \equiv m\omega_0$, where m was an integer. The integral will be zero for any ω that is not equal to $2\pi n/P$, a more complete explanation can be seen in Section 6.2.2.3 [Fourier series], page 71. Therefore, while in the spatial domain the impulses had spacings of P (meters for example), in the frequency space, the spacings between the different impulses are $2\pi/P$ cycles per meters.

6.2.2.6 Convolution theorem

The convolution (shown with the $*$ operator) of the two functions $f(l)$ and $h(l)$ is defined as:

$$c(l) \equiv [f*h](l) = \int_{-\infty}^{\infty} f(\tau)h(l - \tau)d\tau$$

See Section 6.2.1.1 [Convolution process], page 66 for a more detailed physical (pixel based) interpretation of this definition. The fourier transform of convolution ($C(\omega)$) can be written as:

$$C(\omega) = \int_{-\infty}^{\infty}[f*h](l)e^{-i\omega l}dl = \int_{-\infty}^{\infty} f(\tau)\left[\int_{-\infty}^{\infty} h(l - \tau)e^{-i\omega l}dl\right]d\tau$$

To solve the inner integral, lets define $s \equiv l - \tau$, so that $ds = dl$ and $l = s + \tau$ then the inner integral becomes:

$$\int_{-\infty}^{\infty} h(l - \tau)e^{-i\omega l}dl = \int_{-\infty}^{\infty} h(s)e^{-i\omega(s+\tau)}ds = e^{-i\omega\tau}\int_{-\infty}^{\infty} h(s)e^{-i\omega s}ds = H(\omega)e^{-i\omega\tau}$$

where $H(\omega)$ is the Fourier transform of $h(l)$. Substituting this result for the inner integral above, we get:

$$C(\omega) = H(\omega) \int_{-\infty}^{\infty} f(\tau)e^{-i\omega\tau}d\tau = H(\omega)F(\omega) = F(\omega)H(\omega)$$

where $F(\omega)$ is the Fourier transform of $f(l)$. So multiplying the Fourier transform of two functions individually, we get the Fourier transform of their convolution. The convolution theorem also proves a relation between the convolutions in the frequency space. Let's define:

$$D(\omega) \equiv F(\omega)*H(\omega)$$

Applying the inverse Fourier Transform or synthesis equation (Section 6.2.2.4 [Fourier transform], page 73) to both sides and following the same steps above, we get:

$$d(l) = f(l)h(l)$$

Where $d(l)$ is the inverse Fourier transform of $D(\omega)$. We can therefore re-write the two equations above formally as the convolution theorem:

$$\mathcal{F}[f*h] = \mathcal{F}[f]\mathcal{F}[h]$$

$$\mathcal{F}[fh] = \mathcal{F}[f] * \mathcal{F}[h]$$

Besides its usefulness in blurring an image by convolving it with a given kernel, the convolution theorem also enables us to do another very useful operation in data analysis: to match the blur (or PSF) between two images taken with different telescopes/cameras or under different atmospheric conditions. This process is also known as de-convolution. Let's take $f(l)$ as the image with a narrower PSF (less blurry) and $c(l)$ as the image with a wider PSF which appears more blurred. Also let's take $h(l)$ to represent the kernel that should be convolved with the sharper image to create the more blurry image. Above, we proved the relation between these three images through the convolution theorem. But there, we assumed that $f(l)$ and $h(l)$ are known (given) and the convolved image is desired.

In deconvolution, we have $f(l)$ –the sharper image– and $f*h(l)$ –the more blurry image– and we want to find the kernel $h(l)$. The solution is a direct result of the convolution theorem:

$$\mathcal{F}[h] = \frac{\mathcal{F}[f*h]}{\mathcal{F}[f]} \qquad \text{or} \qquad h(l) = \mathcal{F}^{-1}\left[\frac{\mathcal{F}[f*h]}{\mathcal{F}[f]}\right]$$

While this works really nice, it has two problems:

- If $\mathcal{F}[f]$ has any zero values, then the inverse Fourier transform will not be a number!
- If there is significant noise in the image, then the high frequencies of the noise are going to significantly reduce the quanlity of the final result.

A standard solution to both these problems is the Weiner deconovolution algorithm[13].

6.2.2.7 Sampling theorem

Our mathematical functions are continuous, however, our data collecting and measuring tools are discrete. Here we want to give a mathematical formulation for digitizing the continuous mathematical functions so that later, we can retrieve the continuous function from the digitized recorded input. Assuming that we have a continuous function $f(l)$, then we can define $f_s(l)$ as the 'sampled' $f(l)$ through the Dirac comb (see Section 6.2.2.5 [Dirac delta and comb], page 74):

$$f_s(l) = f(l)\mathrm{III}_P = \sum_{n=-\infty}^{\infty} f(l)\delta(l - nP)$$

The discrete data-element f_k (for example, a pixel in an image), where k is an integer, can thus be represented as:

$$f_k = \int_{-\infty}^{\infty} f_s(l)dl = \int_{-\infty}^{\infty} f(l)\delta(l - kP)dt = f(kP)$$

Note that in practice, our discrete data points are not found in this fashion. Each detector pixel (in an image for example) has an area and averages the signal it receives over that area, not a mathematical point as the Dirac δ function defines. However, as long as the variation in the signal over one detector pixel is not significant, this can be a good approximation. Having put this issue to the side, we can now try to find the relation between the fourier transforms of the unsampled $f(l)$ and the sampled $f_s(l)$. For a more clear notation, let's define:

$$F_s(\omega) \equiv \mathcal{F}[f_s]$$

$$D(\omega) \equiv \mathcal{F}[\mathrm{III}_P]$$

Then using the Convolution theorem (see Section 6.2.2.6 [Convolution theorem], page 75), $F_s(\omega)$ can be written as:

$$F_s(\omega) = \mathcal{F}[f(l)\mathrm{III}_P] = F(\omega){*}D(\omega)$$

Finally, from the definition of convolution and the Fourier transform of the Dirac comb (see Section 6.2.2.5 [Dirac delta and comb], page 74), we get:

---

[13] https://en.wikipedia.org/wiki/Wiener_deconvolution

$$F_s(\omega) = \int_{-\infty}^{\infty} F(\omega)D(\omega - \mu)d\mu$$

$$= \frac{1}{P} \sum_{n=-\infty}^{\infty} \int_{-\infty}^{\infty} F(\omega)\delta\left(\omega - \mu - \frac{2\pi n}{P}\right)d\mu$$

$$= \frac{1}{P} \sum_{n=-\infty}^{\infty} F\left(\omega - \frac{2\pi n}{P}\right).$$

$F(\omega)$ was only a simple function, see Figure 6.3(left). However, from the sampled Fourier transform function we see that $F_s(\omega)$ is the superposition of infinite copies of $F(\omega)$ that have been shifted, see Figure 6.3(right). From the equation, it is clear that the shift in each copy is $2\pi/P$.

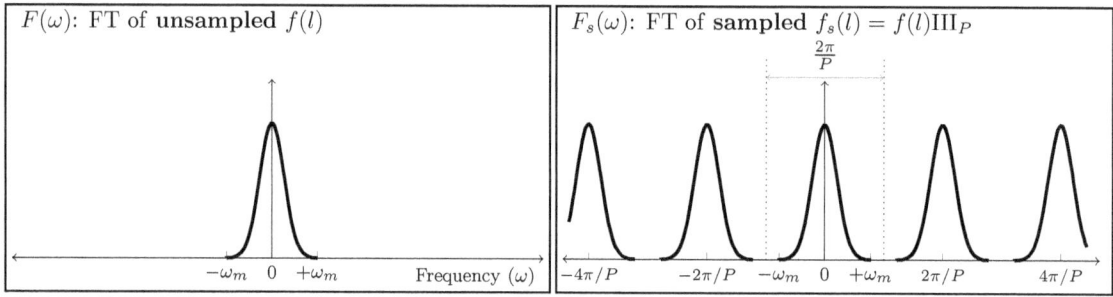

Figure 6.3: Sampling causes infinite repetition in the frequency domain. FT is an abbreviation for 'Fourier transform'. ω_m represents the maximum frequency present in the input. $F(\omega)$ is only symmetric on both sides of 0 when the input is real (not complex). In general $F(\omega)$ is complex and thus cannot be simply plotted like this. Here we have assumed a real Gaussian $f(t)$ which has produced a Gaussian $F(\omega)$.

The input $f(l)$ can have any distribution of frequencies in it. In the example of Figure 6.3(left), the input consisted of a range of frequencies equal to $\Delta\omega = 2\omega_m$. Fortunately as Figure 6.3(right) shows, the assumed pixel size (P) we used to sample this hypothetical function was such that $2\pi/P > \Delta\omega$. The consequence is that each copy of $F(\omega)$ has become completely separate from the surrounding copies. Such a digitized (sampled) dataset is thus called *over-sampled*. When $2\pi/P = \Delta\omega$, P is just small enough to finely separte even the largest frequencies in the input signal and thus it is known as *critically-sampled*. Finally if $2\pi/P < \Delta\omega$ we are dealing with an *under-sampled* dataset. In an under-sampled dataset, the separate copies of $F(\omega)$ are going to overlap and this will deprive us of recovering high constituent frequencies of $f(l)$. The effects of under-sampling in an image with high rates of change (for example a brick wall imaged from a distance) can clearly be visually seen and is known as *aliasing*.

When the input $f(l)$ is composed of a finite range of frequencies, $f(l)$ is known as a *band-limited* function. The example in Figure 6.3(left) was a nice demonstration of such a case: for all $\omega < -\omega_m$ or $\omega > \omega_m$, we have $F(\omega) = 0$. Therefore, when the input function is band-limited and our detector's pixels are placed such that we have critically (or over-) sampled it, then we can exactly reproduce the continuous $f(l)$ from the discrete or digitized samples. To do that, we just have to isolate one copy of $F(\omega)$ from the infinite copies and take its inverse Fourier transform.

This ability to exactly reproduce the continuous input from the sampled or digitized data leads us to the *sampling theorem* which connects the inherent property of the continuous signal (its maximum frequency) to that of the detector (the spacings between its pixels). The sampling theorem states that the full (continuous) signal can be recovered when the pixel size (P) and the maximum constituent frequency in the signal (ω_m) have the following relation[14]:

$$\frac{2\pi}{P} > 2\omega_m$$

This relation was first formulated by Harry Nyquist (1889 – 1976 A.D.) in 1928 and formally proved in 1949 by Claude E. Shannon (1916 – 2001 A.D.) in what is now known as the Nyquist-Shannon sampling theorem. In signal processing, the signal is produced (synthesized) by a transmitter and is received and de-coded (analyzed) by a receiver. Therefore producing a band-limited signal is necessary.

In astronomy, we do not produce the shapes of our targets, we are only observers. Galaxies can have any shape and size, therefore ideally, our signal is not band-limited. However, since we are always confined to observing through an aperture, the aperture will cause a point source (for which $\omega_m = \infty$) to be spread over several pixels. This spread is quantitatively known as the point spread function or PSF. This spread does blur the image which is undesirable; however, for this analysis it produces the positive outcome that there will be a finite ω_m. Though we should caution that any detector will have noise which will add lots of very high frequency (ideally inifinite) changes between the pixels. However, the coefficients of those noise frequencies are usually exceedingly small.

6.2.2.8 Discrete Fourier transform

As we have stated several times so far, the input image is a digitized, pixelated or discrete array of values ($f_s(l)$, see Section 6.2.2.7 [Sampling theorem], page 77). The input is not a continuous function. Also, all our numerical calculations can only be done on a sampled, or discrete Fourier transform. Note that $F_s(\omega)$ is not discrete, it is continuous. One way would be to find the analytic $F_s(\omega)$, then sample it at any desired "freq-pixel"[15] spacing. However, this process would involve two steps of operations and computers in particular are not too good at analytic operations for the first step. So here, we will derive a method to directly find the 'freq-pixel'ated $F_s(\omega)$ from the pixelated $f_s(l)$. Let's start with the definition of the Fourier transform (see Section 6.2.2.4 [Fourier transform], page 73):

$$F_s(\omega) = \int_{-\infty}^{\infty} f_s(l)e^{-i\omega l}dl$$

From the defintion of $f_s(\omega)$ (using x instead of n) we get:

---

[14] This equation is also shown in some places without the 2π. Whether 2π is included or not depends on how you define the frequency

[15] We are using the made-up word "freq-pixel" so they are not confused with spatial domain "pixels".

$$F_s(\omega) = \sum_{x=-\infty}^{\infty} \int_{-\infty}^{\infty} f(l)\delta(l - xP)e^{-i\omega l}dl$$

$$= \sum_{x=-\infty}^{\infty} f_x e^{-i\omega xP}$$

Where f_x is the value of $f(l)$ on the point x or the value of the xth pixel. As shown in Section 6.2.2.7 [Sampling theorem], page 77 this function is infinitely periodic with a period of $2\pi/P$. So all we need is the values within one period: $0 < \omega < 2\pi/P$, see Figure 6.3. We want X samples within this interval, so the frequency difference between each frequency sample or freq-pixel is $1/XP$. Hence we will evaluate the equation above on the points at:

$$\omega = \frac{u}{XP} \qquad u = 0, 1, 2, ..., X - 1$$

Therefore the value of the freq-pixel u in the frequency domain is:

$$F_u = \sum_{x=0}^{X-1} f_x e^{-i\frac{ux}{X}}$$

Therefore, we see that for each freq-pixel in the frequency domain, we are going to need all the pixels in the spatial domain[16]. If the input (spatial) pixel row is also X pixels wide, then we can exactly recover the xth pixel with the following summation:

$$f_x = \frac{1}{X} \sum_{u=0}^{X-1} F_u e^{i\frac{ux}{X}}$$

When the input pixel row (we are still only working on 1D data) has X pixels, then it is $L = XP$ spatial units wide. L, or the length of the input data was defined in Section 6.2.2.3 [Fourier series], page 71 and P or the space between the pixels in the input was defined in Section 6.2.2.5 [Dirac delta and comb], page 74. As we saw in Section 6.2.2.7 [Sampling theorem], page 77, the input (spatial) pixel spacing (P) specifies the range of frequencies that can be studied and in Section 6.2.2.3 [Fourier series], page 71 we saw that the length of the (spatial) input, (L) determines the resolution (or size of the freq-pixels) in our discrete fourier transformed image. Both result from the fact that the frequency domain is the inverse of the spatial domain.

6.2.2.9 Fourier operations in two dimensions

Once all the relations in the previous sections have been clearly understood in one dimension, it is very easy to generalize them to two or even more dimentions since each dimension is by definition independent. Previously we defined l as the continuous variable in 1D and the inverse of the period in its direction to be ω. Lets show the second spatial direction with

---

[16] So even if one pixel is a blank pixel (see Section 6.1.3 [Blank pixels], page 60), all the pixels in the frequency domain will also be blank.

m the the inverse of the period in the second dimension with ν. The Fourier transform in 2D (see Section 6.2.2.4 [Fourier transform], page 73) can be written as:

$$F(\omega, \nu) = \int_{-\infty}^{\infty} \int_{-\infty}^{\infty} f(l, m) e^{-i(\omega l + \nu m)} dl$$

$$f(l, m) = \int_{-\infty}^{\infty} \int_{-\infty}^{\infty} F(\omega, \nu) e^{i(\omega l + \nu m)} dl$$

The 2D Dirac $\delta(l, m)$ is non-zero only when $l = m = 0$. The 2D Dirac comb (or Dirac brush! See Section 6.2.2.5 [Dirac delta and comb], page 74) can be written in units of the 2D Dirac δ. For most image detectors, the sides of a pixel are equal in both dimentions. So P remains unchanged, if a specific device is used which has non-square pixels, then for each dimention a different value should be used.

$$\text{III}_P(l, m) \equiv \sum_{j=-\infty}^{\infty} \sum_{k=-\infty}^{\infty} \delta(l - jP, m - kP)$$

The Two dimensional Sampling theorem (see Section 6.2.2.7 [Sampling theorem], page 77) is thus very easily derived as before since the frequencies in each dimention are independent. Let's take ν_m as the maximum frequency along the second dimention. Therefore the two dimensional sampling theorem says that a 2D band-limited function can be recovered when the following conditions hold[17]:

$$\frac{2\pi}{P} > 2\omega_m \qquad \text{and} \qquad \frac{2\pi}{P} > 2\nu_m$$

Finally, let's represent the pixel counter on the second dimension in the spatial and frequency domains with y and v respectively. Also lets assume that the input image has Y pixels on the second dimension. Then the two dimensional discrete Fourier transform and its inverse (see Section 6.2.2.8 [Discrete Fourier transform], page 79) can be written as:

$$F_{u,v} = \sum_{x=0}^{X-1} \sum_{y=0}^{Y-1} f_{x,y} e^{-i(\frac{ux}{X} + \frac{vy}{Y})}$$

$$f_{x,y} = \frac{1}{XY} \sum_{u=0}^{X-1} \sum_{v=0}^{Y-1} F_{u,v} e^{i(\frac{ux}{X} + \frac{vy}{Y})}$$

---

[17] If the pixels are not a square, then each dimention has to use the respective pixel size, but since most imagers have square pixels, we assume so here too

6.2.2.10 Edges in the frequency domain

With a good grasp of the frequency domain, we can revist the problem of convolution on the image edges, see Section 6.2.1.2 [Edges in the spatial domain], page 67. When we apply the convolution theorem (see Section 6.2.2.6 [Convolution theorem], page 75) to convolve an image, we first take the discrete Fourier transforms (DFT, Section 6.2.2.8 [Discrete Fourier transform], page 79) of both the input image and the kernel, then we multiply them with each other and then take the inverse DFT to construct the convolved image. Ofcourse, in order to multiply them with each other in the frequency domain, the two images have to be the same size, so let's assume that we pad the kernel (it is usually smaller than the input image) with zero valued pixels in both dimensions so it becomes the same size as the input image before the DFT.

Having multiplied the two DFTs, we now apply the inverse DFT which is where the problem is usually created. If the DFT of the kernel only had values of 1 (unrealistic condition!) then there would be no problem and the inverse DFT of the multiplication would be identical with the input. However in real situations, the kernel's DFT has a maximum of 1 (because the sum of the kernel has to be one, see Section 6.2.1.1 [Convolution process], page 66) and decreases something like the hypothetical profile of Figure 6.3. So when multiplied with the input image's DFT, the coefficients or magnitudes (see Section 6.2.2.2 [Circles and the complex plane], page 70) of the smallest frequency (or the sum of the input image pixels) remains unchanged, while the magnitudes of the higher frequencies are significantly reduced.

As we saw in Section 6.2.2.7 [Sampling theorem], page 77, the Fourier tranform of a discrete input will be infinitely repeated. In the final inverse DFT step, the input is in the frequency domain (the multiplied DFT of the input image and the kernel DFT). So the result (our output convolved image) will be infinitely repeated in the spatial domain. In order to accurately reconstruct the input image, we need all the frequencies with the correct magnitudes. However, when the magnitudes of higher frequencies are decreased, longer periods (shorter frequencies) will dominate in the reconstructed pixel values. Therefore, when constructing a pixel on the edge of the image, the newly empowered longer periods will look beyond the input image edges and will find the repeated input image there. So if you convolve an image in this fashion using the convolution theorem, when a bright object exists on one edge of the image, its blurred wings will be present on the other side of the convolved image. This is often termed as circular convolution or cyclic convolution.

So as long as we are dealing with convolution in the frequency domain, there is nothing we can do about the image edges. The least we can do is to eliminate the ghosts of the other side of the image. So, we add zero valued pixels to both the input image and the kernel in both dimentions so the image that will be covolved has the a size equal to the sum of both images in each dimention. Ofcourse, the effect of this zero-padding is that the sides of the output convolved image will become dark. To put it another way, the edges are going to drain the flux from nearby objects. But at least it is consistent across all the edges of the image and is predictable. In Convolve, you can see the padded images when inspecting the frequency domain convolution steps with the `--viewfreqsteps` option.

6.2.3 Spatial vs. Frequency domain

With the discussions above it might not be clear when to choose the spatial domain and when to choose the frequency domain. Here we will try to list the benefits of each.

The spatial domain,

- Can correct for the edge effects of convolution, see Section 6.2.1.2 [Edges in the spatial domain], page 67.

- Can operate on blank pixels.

- Can be faster than frequency domain when the kernel is small (in terms of the number of pixels on the sides).

The frequency domain,

- Will be much faster when the image and kernel are both large.

As a general rule of thumb, when working on an image of modelled profiles use the frequency domain and when working on an image of real (observed) objects use the spatial domain (corrected for the edges). The reason is that if you apply a frequency domain convolution to a real image, you are going to loose information on the edges and generally you don't want large kernels. But when you have made the profiles in the image your self, you can just make a larger input image and crop the central parts to completely remove the edge effect, see Section 8.1.2 [If convolving afterwards], page 134. Also due to oversampling, both the kernels and the images can become very large and the speed boost of frequency domain convolution will significantly improve the processing time, see Section 8.1.1.6 [Oversampling], page 134.

6.2.4 Convolution kernel

All the programs that need convolution will need to be given a convolution kernel file and extension. In most cases (other than Convolve, see Section 6.2 [Convolve], page 65) the kernel file name is optional. However, the extension is necessary and must be specified either on the command line or at least one of the configuration files (see Section 4.2 [Configuration files], page 37). Within Gnuastro, there are two ways to create a kernel image:

- MakeProfiles: You can use MakeProfiles to create a parameteric (based on a radial function) kernel, see Section 8.1 [MakeProfiles], page 129. By default MakeProfiles will make the Gaussian and Moffat profiles in a separate file so you can feed it into any of the programs.

- ConvertType: You can write your own desired kernel into a text file table and convert it to a FITS file with ConvertType, see Section 5.2 [ConvertType], page 51. Just be careful that the kernel has to have an odd number of pixels along its two axises, see Section 6.2.1.1 [Convolution process], page 66. All the programs that do convolution will normalize the kernel internally, so if you choose this option, you don't have to worry about normalizing the kernel. Only within Convolve, there is an option to disable normalization, see Section 6.2.5 [Invoking Convolve], page 84.

The two options to specify a kernel file name and its extension are shown below. These are common between all the programs that will do convolution.

-k

--kernel   (=STR) The convolution kernel file name. The BITPIX (data type) value of this file can be any standard type and it does not necessarily have to be normalized. Several operations will be done on the kernel image prior to the program's processing:

- It will be converted to floating point type.

- All blank pixels (see Section 6.1.3 [Blank pixels], page 60) will be set to zero.

- It will be normalized so the sum of its pixels equal unity.

- It will be flipped so the convolved image has the same orientation. This is only relevant if the kernel is not circular. See Section 6.2.1.1 [Convolution process], page 66.

-U

--khdu   (=STR) The convolution kernel HDU. Although the kernel file name is optional, before running any of the programs, they need to have a value for --khdu even if the default kernel is to be used. So be sure to keep its value in at least one of the configuration files (see Section 4.2 [Configuration files], page 37). By default, the system configuration file has a value.

--fullconvolution
            Ignore the (possible) channels in the mesh grid when doing spatial convolution, see Section 6.4.2 [Tiling an image], page 95. When applied over a mesh grid, spatial convolution will be done independently on each channel. This is necessary when the noise properties of each channel are different and so the pixels should not be mixed. With this option, all channel information is going to be ignored. Currently this option is not deployed for the frequency space convolutions.

6.2.5 Invoking Convolve

Convolve an input image with a known kernel. The general template for convolve is:

```
$ astconvolve [OPTION...] ASTRdata
```

One line examples:

```
$ astconvolve --kernel=psf.fits mockimg.fits
$ astconvolve --kernel=sharperimage.fits --makekernel\
              blurryimage.fits
```

The only argument accepted by Convolve is an input image file. Some of the options are the same between Convolve and some other Gnuastro programs. Therefore, to avoid repetition, they will not be repeated here. For the full list of options shared by all Gnuastro programs, please see Section 4.1.4 [Common options], page 34. Section 6.4.2.4 [Mesh grid options], page 100 lists all the options related to spefiying a mesh grid which is currently only used in spatial convolution. Note that here, no interpolation or smoothing is defined, only channels and the mesh size are important. Section 6.2.4 [Convolution kernel], page 83 lists the the convolution kernel options.

It is also possible to specify a mask image for the input. In that case, see Section 6.4.3 [Mask image], page 102. Here we will only explain the options particular to Convolve. Run Convolve with --help in order to see the full list of options Convolve accepts, irrespective of where they are explained in this manual.

--nokernelflip
            Do not flip the kernel after reading it the spatial domain convolution. This can be useful if the flipping has already been applied to the kernel.

`--nokernelnorm`

> Do not normalize the kernel after reading it, such that the sum of its pixels is unity.

`-f`

`--frequency`

> Convolve using discrete fourier transform in the frequency domain: The fourier transform of both arrays is first calculated and multiplied. Then the inverse fourier transform is applied to the product to give the final convolved image.
>
> For large images, this process will be more efficient than convolving in the spatial domain. However, the edges of the image will loose some flux, see Section 6.2.1.2 [Edges in the spatial domain], page 67.

`-p`

`--spatial`

> Convolve in the spatial domain, see Section 6.2.1.1 [Convolution process], page 66.

`--viewfreqsteps`

> With this option a file with the initial name of the output file will be created that is suffixed with `_freqsteps.fits`, all the steps done to arrive at the final convolved image are saved as extensions in this file. The extensions in order are:
>
> 1. The padded input image. In frequency domain convolution the two images (input and convolved) have to be the same size and both should be padded by zeros.
>
> 2. The padded kernel, similar to the above.
>
> 3. The Fourier spectrum of the forward Fourier transform of the input image. Note that the fourier transform is a complex operation (and not viewable in one image!) So we either have to show the 'Fourier spectrum' or the 'Phase angle'. For the complex number $a + ib$, the Fourier spectrum is defined as $\sqrt{a^2 + b^2}$ while the phase angle is defined as $arctan(b/a)$.
>
> 4. The Fourier spectrum of the forward Fourier transform of the kernel image.
>
> 5. The Fourier spectrum of the multiplied (through complex arithmetic) transformed images.
>
> 6. The inverse Fourier transform of the multiplied image. If you open it, you will see that the convolved image is now in the center, not on one side of the image as it started with (in the padded image of the first extension). If you are working on a mock image which originally had pixels of precisely 0.0, you will notice that in those parts that your convolved profile(s) did not conver, the values are now $\sim 10^{-18}$, this is due to floating-point round off errors. Therefore in the final step (when cropping the central parts of the image), we also remove any pixel with a value less than 10^{-17}.

`-m`

`--makekernel`

> (`INT`) If this option is called, Convolve will do deconvolution (see Section 6.2.2.6 [Convolution theorem], page 75). The image specified by the `--kernel` option

is assumed to be the sharper (less blurry) image and the input image is assumed to be the more blurry image. The two images have to be the same size. Some notes to take into account for a good result:

Noise has large frequencies which can make the result less reliable for the higher frequencies of the kernel. So all the frequencies which have a spectrum smaller than 0.005 in the frequency domain are set to zero and not divided. This will cause the wings of the final kernel to be flatter than they would ideally be which will make the convolved image result unreliable. The value given to this option will be used as the maximum radius of the kernel. Any pixel in the final kernel that is larger than this distance from the center will be set to zero.

- Choose a bright (unsaturated) star and use a region box (with ImageCrop for example, see Section 6.1 [ImageCrop], page 58) that is sufficiently above the noise.

- Use ImageWarp (see Section 6.3 [ImageWarp], page 86) to warp the pixel grid so the star's center is exactly on the center of the central pixel in the cropped image. This will certainly slightly degrade the result, however, it is necessary. If there are multiple good stars, you can shift all of them, then normalize them (so the sum of each star's pixels is one) and then take their average to decrease this effect.

- The shifting might move the center of the star by one pixel in any direction, so crop the central pixel of the warped image to have a clean image for the deconvolution.

6.3 ImageWarp

Image warpring is the process of mapping the pixels of one image onto a new pixel grid. This process is sometimes known as transformation, however following the discussion of Heckbert 1989[18] we will not be using that term because it can be confused with only pixel value or flux transformations. Here we specifically mean the pixel grid transformation which is better conveyed with 'warp'.

Image wrapping is a very important step in astronomy, both in observational data analysis and in simulating modeled images. In modelling, warping an image is necessary when we want to apply grid transformations to the initial models, for example in simulating gravitational lensing (Radial warpings are not yet included in ImageWarp). Observational reasons for warping an image are listed below:

- **Noise:** Most scientifically interesting targets are inherently faint (have a very low Signal to noise ratio). Therefore one short exposure is not enough to detect such objects that are drowned deeply in the noise. We need multiple exposures so we can add them together and increase the objects' signal to noise ratio. Keeping the telescope fixed on one field of the sky is practically impossible. Therefore very deep observations have to put into the same grid before adding them.

- **Resolution:** If we have multiple images of one patch of the sky (hopefully at multiple orientations) we can warp them to the same grid. The multiple orientations will allow us

---

[18] Paul S. Heckbert. 1989. *Fundamentals of Texture mapping and Image Warping*, Master's thesis at University of California, Berkely.

to 'guess' the values of pixels on an output pixel grid that has smaller pixel sizes and thus increase the resolution of the output. This process of merging multiple observations is known as Mosaicing.

- **Cosmic rays:** Cosmic rays can randomly fall on any part of an image. If they collide vertically with the camera, they are going to create a very sharp and bright spot that in most cases can be separated easily[19]. However, depending on the depth of the camera pixels, and the angle that a cosmic rays collides with it, it can cover a line-like larger area on the CCD which makes the detection using their sharp edges very hard and error prone. One of the best methods to remove cosmic rays is to compare multiple images of the same field. To do that, we need all the images to be on the same pixel grid.

- **Optical distortion:** (Not yet included in ImageWarp) In wide field images, the optical distortion that occurs on the outer parts of the focal plane will make accurate comparison of the objects at various locations impossible. It is therefore necessary to warp the image and correct for those distortions prior to the analysis.

- **Detector not on focal plane:** In some cases (like the Hubble Space Telescope ACS and WFC3 cameras), the CCD might be tilted compared to the focal plane, therefore the recorded CCD pixels have to be projected onto the focal plane before further analysis.

6.3.1 Warping basics

Let's take $\begin{bmatrix} u & v \end{bmatrix}$ as the coordinates of a point in the input image and $\begin{bmatrix} x & y \end{bmatrix}$ as the coordinates of that same point in the output image[20]. The simplest form of coordinate transformation (or warping) is the scaling of the coordinates, let's assume we want to scale the first axis by M and the second by N, the output coordinates of that point can be calculated by

$$\begin{bmatrix} x \\ y \end{bmatrix} = \begin{bmatrix} Mu \\ Nv \end{bmatrix} = \begin{bmatrix} M & 0 \\ 0 & N \end{bmatrix} \begin{bmatrix} u \\ v \end{bmatrix}$$

Note that these are matrix multiplications. We thus see that we can represent any such grid warping as a matrix. Another thing we can do with this 2×2 matrix is to rotate the output coordinate around the common center of both coordinates. If the output is rotated anticlockwise by θ degrees from the positive (to the right) horizontal axis, then the warping matrix should become:

$$\begin{bmatrix} x \\ y \end{bmatrix} = \begin{bmatrix} u\cos\theta - v\sin\theta \\ u\sin\theta + v\cos\theta \end{bmatrix} = \begin{bmatrix} \cos\theta & -\sin\theta \\ \sin\theta & \cos\theta \end{bmatrix} \begin{bmatrix} u \\ v \end{bmatrix}$$

We can also flip the coordinates around the first axis, the second axis and the coordinate center with the following three matrices respectively:

$$\begin{bmatrix} 1 & 0 \\ 0 & -1 \end{bmatrix} \quad \begin{bmatrix} -1 & 0 \\ 0 & 1 \end{bmatrix} \quad \begin{bmatrix} -1 & 0 \\ 0 & -1 \end{bmatrix}$$

---

[19] All astronomical targets are blurred with the PSF, see Section 8.1.1.2 [Point Spread function], page 130, however a cosmic ray is not and so it is very sharp (it suddenly stops at one pixel).

[20] These can be any real number, we are not necessarily talking about integer pixels here.

The final thing we can do with this definition of a 2×2 warping matrix is shear. If we want the output to be sheared along the first axis with A and along the second with B, then we can use the matrix:

$$\begin{bmatrix} 1 & A \\ B & 1 \end{bmatrix}$$

To have one matrix representing any combination of these steps, you use matrix multiplication, see Section 6.3.2 [Merging multiple warpings], page 89. So any combinations of these transformations can be displayed with one 2×2 matrix:

$$\begin{bmatrix} a & b \\ c & d \end{bmatrix}$$

The transformations above can cover a lot of the needs of most coordinate transformations. However they are limited to mapping the point $\begin{bmatrix} 0 & 0 \end{bmatrix}$ to $\begin{bmatrix} 0 & 0 \end{bmatrix}$. Therefore they are useless if you want one coordinate to be shifted compared to the other one. They are also space invariant, meaning that all the coordinates in the image will recieve the same transformation. In other words, all the pixels in the output image will have the same area if placed over the input image. So transformations which require varying output pixel sizes like projections cannot be applied through this 2×2 matrix either (for example for the tilted ACS and WFC3 camera detectors on board the Hubble space telescope).

To add these further capabilities, namely translation and projection, we use the homogeneous coordinates. They were defined about 200 years ago by August Ferdinand Möbius (1790 – 1868). For simplicity, we will only discuss points on a 2D plane and avoid the complexities of higher dimensions. We cannot provide a deep mathematical introduction here, interested readers can get a more detailed explanation from Wikipedia[21] and the references therein.

By adding an extra coordinate to a point we can add the flexibility we need. The point $\begin{bmatrix} x & y \end{bmatrix}$ can be represented as $\begin{bmatrix} xZ & yZ & Z \end{bmatrix}$ in homogeneous coordinates. Therefore multiplying all the coordinates of a point in the homogenous coordinates with a constant will give the same point. Put another way, the point $\begin{bmatrix} x & y & Z \end{bmatrix}$ corresponds to the point $\begin{bmatrix} x/Z & y/Z \end{bmatrix}$ on the constant Z plane. Setting $Z = 1$, we get the input image plane, so $\begin{bmatrix} u & v & 1 \end{bmatrix}$ corresponds to $\begin{bmatrix} u & v \end{bmatrix}$. With this definition, the transformations above can be generally written as:

$$\begin{bmatrix} x \\ y \\ 1 \end{bmatrix} = \begin{bmatrix} a & b & 0 \\ c & d & 0 \\ 0 & 0 & 1 \end{bmatrix} \begin{bmatrix} u \\ v \\ 1 \end{bmatrix}$$

We thus acquired 4 extra degrees of freedom. By giving non-zero values to the zero valued elements of the last column we can have translation (try the matrix multiplication!). In

---

[21] http://en.wikipedia.org/wiki/Homogeneous_coordinates

general, any coordinate transformation that is represented by the matrix below is known as an affine transformation[22]:

$$\begin{bmatrix} a & b & c \\ d & e & f \\ 0 & 0 & 1 \end{bmatrix}$$

We can now consider translation, but the affine transform is still spatially invariant. Giving non-zero values to the other two elements in the matrix above gives us the projective transformation or Homography[23] which is the most general type of transformation with the 3×3 matrix:

$$\begin{bmatrix} x' \\ y' \\ w \end{bmatrix} = \begin{bmatrix} a & b & c \\ d & e & f \\ g & h & 1 \end{bmatrix} \begin{bmatrix} u \\ v \\ 1 \end{bmatrix}$$

So the output coordinates can be calculated from:

$$x = \frac{x'}{w} = \frac{au + bv + c}{gu + hv + 1} \qquad\qquad y = \frac{y'}{w} = \frac{du + ev + f}{gu + hv + 1}$$

Thus with homography we can change the sizes of the output pixels on the input plane, giving a 'perspective'-like visual impression. This can be quantitatively seen in the two equations above. When $g = h = 0$, the denominator is independent of u or v and thus we have spatial invariance. Homography preserves lines at all orientations. A very useful fact about homography is that its inverse is also a homography. These two properties play a very important role in the implementation of this transformation. A short but instructive and illustrated review of affine, projective and also bilinear mappings is provided in Heckbert 1989[24].

6.3.2 Merging multiple warpings

In Section 6.3.1 [Warping basics], page 87 we saw how one basic warping/transformation can be represented with a 3 by 3 matrix. To make more complex warpings these matrices have to be multiplied through matrix multiplication. However matrix multiplication is not commutative, so the order of the set of matrices you use for the multiplication is going to be very important.

The first warping should be placed as the left-most matrix. The second warping to the right of that and so on. The second transformation is going to occur on the warped coordinates of the first. As an example for merging a few transforms into one matrix, the multiplication below represents the rotation of an image about a point $[\,U \quad V\,]$ anticlockwise from the horizontal axis by an angle of θ. To do this, first we take the origin to $[\,U \quad V\,]$

---

[22] http://en.wikipedia.org/wiki/Affine_transformation

[23] http://en.wikipedia.org/wiki/Homography

[24] Paul S. Heckbert. 1989. *Fundamentals of Texture mapping and Image Warping*, Master's thesis at University of California, Berkely. Note that since points are defined as row vectors there, the matrix is the transpose of the one discussed here.

through translation. Then we rotate the image, then we translate it back to where it was initially. These three operations can be merged in one operation by calculating the matrix multiplication below:

$$\begin{bmatrix} 1 & 0 & U \\ 0 & 1 & V \\ 0 & 0 & 1 \end{bmatrix} \begin{bmatrix} cos\theta & -sin\theta & 0 \\ sin\theta & cos\theta & 0 \\ 0 & 0 & 1 \end{bmatrix} \begin{bmatrix} 1 & 0 & -U \\ 0 & 1 & -V \\ 0 & 0 & 1 \end{bmatrix}$$

6.3.3 Resampling

A digital image is composed of discrete 'picture elements' or 'pixels'. When a real image is created from a camera or detector, each pixel's area is used to store the number of photoelectrons that were created when incident photons collided with that pixel's surface area. This process is called the 'sampling' of a continuous or analog data into digital data. When we change the pixel grid of an image or warp it as we defined in Section 6.3.1 [Warping basics], page 87, we have to 'guess' the flux value of each pixel on the new grid based on the old grid, or resample it. Because of the 'guessing', any form of warping on the data is going to degrade the image and mix the original pixel values with each other. So if an analysis can be done on an un-warped data image, it is best to leave the image untouched and pursue the analysis. However as discussed in Section 6.3 [ImageWarp], page 86 this is not possible most of the times, so we have to accept the problem and re-sample the image.

In most applications of image processing, it is sufficient to consider each pixel to be a point and not an area. This assumption can significantly speed up the processing of an image and also the simplicity of the code. It is a fine assumption when the signal to noise ratio of the objects are very large. The question will then be one of interpolation because you have multiple points distributed over the output image and you want to find the values at the pixel centers. To increase the accuracy, you might also sample more than one point from within a pixel giving you more points for a more accurate interpolation in the output grid.

However, interpolation has several problems. The first one is that it will depend on the type of function you want to assume for the interpolation. For example you can choose a bi-linear or bi-cubic (the 'bi's are for the 2 dimensional nature of the data) interpolation method. For the latter there are various ways to set the constants[25]. Such functional interpolation functions can fail seriously on the edges of an image. They will also need normalization so that the flux of the objects before and after the warpings are comparable. The most basic problem with such techniques is that they are based on a point while a detector pixel is an area. They add a level of subjectivitiy to the data (make more assumptions through the functions than the data can handle). For most applications this is fine, but in scientific applications where detection of the faintest possible galaxies or fainter parts of bright galaxies is our aim, we cannot afford this loss. Because of these reasons ImageWarp will not use such interpolation techniques.

ImageWarp will do interpolation based on "pixel mixing"[26] or "area resampling". This is also what the Hubble Space Telescope pipeline calles "Drizzling"[27]. This technique re-

---

[25] see http://entropymine.com/imageworsener/bicubic/ for a nice introduction.

[26] For a graphic demonstration see http://entropymine.com/imageworsener/pixelmixing/.

[27] http://en.wikipedia.org/wiki/Drizzle_(image_processing)

quires no functions, it is thus non-parametric. It is also the closest we can get (make least assumptions) to what actually happens on the detector pixels. The basic idea is that you reverse-transform each output pixel to find which pixels of the input image it covers and what fraction of the area of the input pixels are covered. To find the output pixel value, you simply sum the value of each input pixel weighted by the overlap fraction (between 0 to 1) of the output pixel and that input pixel. Through this process, pixels are treated as an area not as a point (which is how detectors create the image), also the brightness (see Section 8.1.3 [Flux Brightness and magnitude], page 134) of an object will be left completely unchanged.

If there are very high spatial-frequency signals in the image (for example fringes) which vary on a scale smaller than your output image pixel size, pixel mixing can cause ailiasing[28]. So if the input image has fringes, they have to be calculated and removed separately (which would naturally be done in any astronomical application). Because of the PSF no astronomical target has a sharp change in the signal so this issue is less important for astronoimcal applications, see Section 8.1.1.2 [Point Spread function], page 130.

6.3.4 Invoking ImageWarp

The general template for invoking ImageWarp is:

```
$ astimgwarp [OPTIONS...] [matrix.txt] InputImage
```

One line examples:

```
$ astimgwarp matrix.txt image.fits
$ astimgwarp --matrix=0.2,0,0.4,0,0.2,0.4,0,0,1 image.fits
$ astimgwarp --matrix="0.7071,-0.7071  0.7071,0.7071" image.fits
```

ImageWarp can accept two arguments, one (the input image) is mandatory if any processing is to be done. An optional argument is a plain text file that will keep the warp/transform matrix, see Section 6.3.1 [Warping basics], page 87. There is also the --matrix option from which the matrix can be literally specified on the command line. If both are present when calling ImageWarp, the contents of the plain text file have higher precedence. The general options to all Gnuastro programs can be seen in Section 4.1.4 [Common options], page 34.

By default the WCS (World Coordinate System) information of the input image is going to be corrected in the output image. WCSLIB will save the input WCS information in the PC matrix[29]. To correct the WCS, ImageWarp multiplies the PC matrix with the inverse of the specified transformation matrix. Also the CRPIX point is going to be changed to its correct place in the output image coordinates. This behavior can be disabled with the --nowcscorrection option.

To be the most accurate the input image will be converted to double precision floating points and all the processing will be done in this format. By default, in the end, the output image will be converted back to the input image data type. Note that if the input type was not a floating point format, then the floating point output pixels are going to be rounded to the nearest integer (using the round function in the C programming language) which can lead to a loss of data. This behavior can be disabled with the --doubletype option. The input file and input warping matrix elements are stored in the output's header.

---

[28] http://en.wikipedia.org/wiki/Aliasing

[29] Greisen E.W., Calabretta M.R. (2002) Representation of world coordinates in FITS. Astronomy and Astrophysics, 395, 1061-1075.

> **Coordinates of pixel center**: Based on the FITS standard, integer values are assigned to the center of a pixel and the coordinate [1.0, 1.0] is the center of the bottom left (first) image pixel. So the point [0.0, 0.0] is half a pixel away (in each axis) from the bottom left vertice of the first pixel[30].

`-m`
`--matrix` (=STR) The warp/transformation matrix. All the elements in this matrix must be separated by any number of space, tab or comma (,) characters. If you want to use the first two, then be sure to wrap the matrix within double quotation marks (") so they are not confused with other arguments on the command line, see Section 4.1.3 [Options], page 32. This also applies to values in the configuration files, see Section 4.2.1 [Configuration file format], page 38. The transformation matrix can be either 2 by 2 or 3 by 3 array, see Section 6.3.1 [Warping basics], page 87.

The determinant of the matrix has to be non-zero and it must not contain any non-number values (for example infinities or NaNs). The elements of the matrix have to be written row by row. So for the general homography matrix of Section 6.3.1 [Warping basics], page 87, it should be called with `--matrix=a,b,c,d,e,f,g,h,1`.

`--hstartwcs`
(=INT) Specify the first header keyword number (line) that should be used to read the WCS information, see the full explantion in Section 6.1.4 [Invoking ImageCrop], page 61.

`--hendwcs`
(=INT) Specify the last header keyword number (line) that should be used to read the WCS information, see the full explantion in Section 6.1.4 [Invoking ImageCrop], page 61.

`-n`
`--nowcscorrection`
Do not correct the WCS information of the input image and save it untouched to the output image.

`-z`
`--zerofornoinput`
Set output pixels which do not correspond to any input to zero. By default they are set to blank pixel values, see Section 6.1.3 [Blank pixels], page 60.

`-b`
`--maxblankfrac`
(=FLT) The maximum fractional area of blank pixels over the output pixel. If an output pixel is covered by blank pixels (see Section 6.1.3 [Blank pixels], page 60) for a larger fraction than the value to this option, the output pixel will be set to a blank pixel its self.

---

[30] So if you want to warp the image relative to the bottom left vertice of the bottom left pixel, you have to shift the warping center by [0.5, 0.5], apply your transform then shift back by [-0.5, -0.5]. Similar to the example in see Section 6.3.2 [Merging multiple warpings], page 89. For example see the one line example above which scales the image by one fifth (0.2). Without this correction (if it was `0.2,0,0,0,0.2,0,0,0,1`), the image would not be correctly scaled.

When the fraction is lower, the sum of non-blank pixel values over that pixel will be multiplied by the inverse of this fraction to correct for its flux and not cause discontinuties on the edges of blank regions. Note that even with this correction, discontinuities (very low non-blank values touching blank regions in the output image) might arise depending on the transformation and the blank pixels. So if there are blank pixels in the image, a good value to this option has to be found for that particular image and warp.

-d

--doubletype

By default the output image is going to have the same type as the input image. If this option is called, the output will be in double precision floating point format irrespective of the input data type. When dealing with integer input formats, this option can be useful in checking the results.

6.4 SubtractSky

Raw astronomical images (and even poorly processed images) don't usually have a uniform 'sky' value over their surface prior to processing, see Section 6.4.1 [Sky value], page 93 for a complete definition of the sky value. However, a uniform sky value over the image is vital for further processing. For ground based images (particularly at longer wavelengths) this can be due to actual variations in the atmosphere. Another cause might be systematic biases in the instrument or prior processing. For example stray light in the telescope/camera or bad flat-fielding or bias subtraction. The latter is a major issue in space based images where the atmosphere is no longer a problem.

SubtractSky is a tool to find the sky value and its standard deviation on a grid over the image. The size of the grid will determine how accurately it can account for gradients in the sky value. Such that if the gradient (change of sky value) is too sharp, a smaller grid size has to be chosen. However the results will be most accurate with a larger grid size.

6.4.1 Sky value

The discussion here is taken from Akhlaghi and Ichikawa (2015)[31]. Let's assume that all instrument defects – bias, dark and flat – have been corrected and the brightness (see Section 8.1.3 [Flux Brightness and magnitude], page 134) of a detected object, O, is desired. The sources of flux on pixel i[32] of the image can be written as follows:

- Contribution from the target object, (O_i).
- Contribution from other detected objects, (D_i).
- Undetected objects or the fainter undetected regions of bright objects, (U_i).
- A cosmic ray, (C_i).
- The background flux, which is defined to be the count if none of the others exists on that pixel, (B_i).

---

[31] See the section on sky in Akhlaghi M., Ichikawa. T. (2015). Astrophysical Journal Supplement Series.

[32] For this analysis the dimentionality of the data (image) is irrelevant. So if the data is an image (2D) with width of w pixels, then a pixel located on column x and row y (where all counting starts from zero and $(0, 0)$ is located on the bottom left corner of the image), would have an index: $i = x + y \times w$.

The total flux in this pixel, T_i, can thus be written as:

$$T_i = B_i + D_i + U_i + C_i + O_i.$$

By definition, D_i is detected and it can be assumed that it is correctly subtracted, so that D_i can be set to zero. There are also methods to detect and remove cosmic rays (for example, van Dokkum (2001)[33]) enabling us to set $C_i = 0$. Note that in practice, D_i and U_i are correlated, because they both directly depend on the detection algorithm and its input parameters. Also note that no detection or cosmic ray removal algorithm is perfect. With these limitations in mind, the observed sky value for this pixel (S_i) can be defined as

$$S_i = B_i + U_i.$$

Therefore, as the detection process (algorithm and input parameters) becomes more accurate, or $U_i \to 0$, the sky value will tend to the background value or $S_i \to B_i$. Therefore, while B_i is an inherent property of the data (pixel in an image), S_i depends on the detection process. Over a group of pixels, for example in an image or part of an image, this equation translates to the average of undetected pixels. With this definition of sky, the object flux in the data can be calculated with

$$T_i = S_i + O_i \quad \to \quad O_i = T_i - S_i.$$

Hence, the more accurately S_i is measured, the more accurately the brightness (sum of pixel values) of the target object can be measured (photometry). Similarly, any under-(over-)estimation in the sky will directly translate to an over(under) estimation of the measured object's brightness. In the fainter outskirts of an object a very small fraction of the photo-electrons in the pixels actually belong to objects. Therefore even a small over estimation of the sky value will result in the loss of a very large portion of most galaxies. Based on the definition above, the sky value is only correctly found when all the detected objects (D_i and C_i) have been removed from the data.

6.4.1.1 Finding the sky value

This technique to find the sky value in a distribution was initially proposed in Akhlaghi and Ichikawa 2015[34].

Note that through the difference of the mode and median we have actually 'detected' data in the distribution. However this detection was only based on the total distribution of the data, not its spatial position in each mesh. So we adhere to the definition of Sky value in Section 6.4.1 [Sky value], page 93. Finding the median is very easy, the main problem is in finding the mode through a robust method. In Appendix C of Akhlaghi and Ichikawa (2015) a new approach to finding the mode in any astronomically relevant distribution is introduced.

---

[33] van Dokkum, P. G. (2001). Publications of the Astronomical Society of the Pacific. 113, 1420.

[34] Akhlaghi M., Ichikawa. T. (2015). Astrophysical Journal Supplement Series.

Even when the mode and median are approximately equal, Cosmic rays can significantly bias the calculation of the average. Even if they are very few. However, usually, Cosmic rays have very sharp boundaries and do not fade away into the noise. Therefore, when the histogram of the distribution is plotted, they are clearly separate from the rest of the data. For example see Figure 15 in Akhlaghi and Ichikawa (2015). In such cases, σ-clipping is a perfect tool to remove the effect of such objects in the average and standard deviation. See Section 7.1.2 [Sigma clipping], page 105 for a complete explanation. So after asserting that the mode and median are approximately equal in a mesh (see Section 6.4.2 [Tiling an image], page 95), convergance-based σ-clipping is also applied before getting the final sky value and its standard deviation for a mesh.

6.4.1.2 Sky value misconceptions

As defined in Section 6.4.1 [Sky value], page 93, the sky value is only accurately defined when the detection algorithm is not significantly reliant on the sky value. In particular its detection threshold. However, most signal-based detection tools[35] used the sky value as a reference to define the detection threshold. So these old techniques had to rely on approximations based on other assumptions about the data. A review of those other techniques can be seen in Appendix A of Akhlaghi and Ichikawa (2015)[36]. Since they were extensively used in astronomical data analysis for several decades, such approximations have given rise to a lot of misconceptions, ambiguities and disagreements about the sky value and how to measure it. As a summary, the major methods used until now were an approximation of the mode of the image pixel distribution and σ-clipping.

- To find the mode of a distribution those methods would either have to assume (or find) a certain probablity density function (PDF) or use the histogram. But the image pixels can have any distribution, and the histogram results are very inaccurate (there is a large dispersion) and depend on bin-widths.

- Another approach was to iteratively clip the brightest pixels in the image (which is known as σ-clipping, since the reference was found from the image mean and its stadard deviation or σ). See Section 7.1.2 [Sigma clipping], page 105 for a complete explanation. The problem with σ-clipping was that real astronomical objects have diffuse and faint wings that penetrate deeply into the noise. So only removing their brightest parts is completely useless in removing the systematic bias an object's fainter parts cause in the sky value.

As discussed in Section 6.4.1 [Sky value], page 93, the sky value can only be correctly defined as the average of undetected pixels. Therefore all such approaches that try to approximate the sky value prior to detection are ultimately poor approximations.

6.4.2 Tiling an image

Some of the programs in Gnuastro will need to divide the pixels in an image into individual tiles or a mesh grid to be able to deal with gradients. In this section we will explain the concept in detail and how the user can check the grid. In the case of SubtractSky, if

---

[35] According to Akhlaghi and Ichikawa (2015), signal-based detection is a detection process that realies heavily on assumptions about the to-be-detected objects. This method was the most heavily used technique prior to the introduction of NoiseChisel in that paper.

[36] Akhlaghi M., Ichikawa. T. (2015). Astrophysical Journal Supplement Series.

the image is completely uniform then one sky value will suffice for the whole image. See Section 6.4.1 [Sky value], page 93 for the definition of the sky value. Unfortunately though, as discussed in Section 6.4 [SubtractSky], page 93, in most images taken with ground or space-based telescopes the sky value is not uniform. So we have to break the image up into small tiles on a mesh grid, assume the sky value is constant over them and find the sky value on those tiles.

Meshes are considered to be a square with a side of `--meshsize` pixels. The best mesh size is directly related to the gradient on the image. In practice we assume that the gradient is not significant over each mesh. So if there is a strong gradient (for example in long wavelength ground based images) or the image is of a crowded area where there isn't too much blank area, you have to choose a smaller mesh size. A larger mesh will give more pixels and so the scatter in the results will be less.

For raw image processing, a simple mesh grid is not sufficient. Raw images are the unprocessed outputs of the camera detectors. Large detectors usually have multiple readout channels each with its own amplifier. For example the Hubble Space Telecope Advanced Camera for Surveys (ACS) has four amplifiers over its full detector area dividing the square field of view to four smaller squares. Ground based image detectors are not exempt, for example each CCD of Subaru Telescope's Hyper Suprime-Cam camera (which has 104 CCDs) has four amplifiers, but they have the same height of the CCD and divide the width by four parts.

The bias current on each amplifier is different, and normaly bias subtraction is not accurately done. So even after subtracting the measured bias current, you can usually still identify identify the boundaries of different amplifiers by eye. See Figure 11(a) in Akhlaghi and Ichikawa (2015) for an example. This results in the final reduced data to have non-uniform amplifier-shaped regions with higher or lower background flux values. Such systematic biases will then propagate to all subsequent measurements we do on the data (for example photometry and subsequent stellar mass and star formation rate measurements in the case of galaxies). Therefore an accurate sky subtraction routine should also be able to account for such biases.

To get an accurate result, the mesh boundaries should be located exactly on the amplifier boundaries. Otherwise, some meshes will contain pixels that have been read from two or four different amplifiers. These meshes are going to give very biased results and the amplifier boundary will still be present after sky subtraction. So we define 'channel's. A channel is an independent mesh grid that covers one amplifier to ensure that the meshes do not pass the amplifier boundary. They can also be used in subsequent steps as the area used to identify nearby neighbors to interpolate and smooth the final grid, see Section 6.4.2.2 [Grid interpolation and smoothing], page 98. The number of channels along each axis can be specified by the user at run time through the command line `--nch1` and `--nch2` options or in the configuration files, see Section 4.2 [Configuration files], page 37. The area of each channel will then be tiled by meshes of the given size and subsequent processing will be done on those meshes. If the image is processed or the detector only has one amplifier, you can set the number of channels in both axises to 1.

Unlike the channel size, that has to be an exact multiple of the image size, the mesh size can be any number. If it is not an exact multiple of the image side, the last (rightest, for the first FITS dimention, and highest for the second when viewed in SAO ds9) mesh will have a different size than the rest. If the remainder of the image size divided by mesh size is larger than a certain fraction (value to `--lastmeshfrac`) of the mesh size along each axis, a new (smaller) mesh will be put there instead of a larger mesh. This is done to avoid the last mesh becoming too large compared to the other meshes in the grid. Generally, it is best practice to choose the mesh size such that the last mesh is only a few (negligible) pixels wider or thinner than the rest.

The final mesh grid can be seen on the image with the `--checkmesh` option that is available to all programs which use the mesh grid for localized operations. When this option is called, a multi-extension FITS file with a `_mesh.fits` suffix will be created along with the outputs, see Section 4.5 [Automatic output], page 43. The first extension will be the input image. For each mesh grid the image produces, there will be a subsequent extension. Each pixel in the grid extensions is labled to the mesh that it is part of. You can flip through the extensions to check the mesh sizes and locations compared to the input image.

6.4.2.1 Quantifying signal in a mesh

Noise is characterized with a fixed background value and a certain distribution. For example, for the Gaussian distribution these two are the mean and standard deviation. When we have absolutely no signal and only noise in a dataset, the mean, median and mode of the distribution are equal within statistical errors and approximately equal to the background value. For the next paragraph, let's assume that the background is subtracted and is zero.

Data always has a positive value and will never become negative, see Figure 1 in Akhlaghi and Ichikawa (2015). Therefore, as data is buried into the noise, the mean, median and mode shift to the positive. The mean is the fastest in this shift. The median is slower since it is defined based on an ordered distribution and so is not affected by a small (less than half) number of outliers. Finally, the mode is the slowest to shift to the positive.

Inversing the argument above provides us with the basis of Gnuastro's algorithm to quantify the presence of signal in a mesh. Namely, when the mode and median of a distribution are approximately equal, we can argue that there is no significant signal in that mesh. So we can consider the image to be made of a grid and use this argument to 'detect' signal in each grid element. The median is defined to be the value of the 0.5 quantile in the image. So the only necessary parameter is the minimum acceptable quantile (smaller than 0.5) for the mode in a mesh that we deem accurate. See Section 6.4.2.4 [Mesh grid options], page 100 for an explanation of the options used to customize this behavior.

Since there is sufficient signal in the mesh to bias the analysis on that mesh, any grid element whose mode quantile is smaller than the minimum acceptable quantile is usually kept with a blank value and no value is given to it. Finally, when all the grid elements have been checked, we can interpolate over all the empty elements and smooth the final result to find the sky value over the full image. See Section 6.4.2.2 [Grid interpolation and smoothing], page 98.

Convolving a dataset (that contains signal and noise), creates a positive skewness in it depending on the fraction of data present in the distribution and also the convolution kernel. See Section 3.1.1 in Akhlaghi and Ichikawa (2015) and Section 6.2.1.1 [Convolution process], page 66. This skewness can be interpreted as an increase in the Signal to noise ratio of the objects burried in the noise. Therefore, to obtain an even better measure of the presence of signal in a mesh, the image can be convolved with a given PSF first. This positive skew will result in more distance between the mode an median thereby enabling a more accurate detection of fainter signal, for example the faint wings of bright stars and galaxies. When convolving over a mesh grid, the pixels in each channel will be treated independently. This can be disabled with the `--fullconvolution` option. See Section 6.2.4 [Convolution kernel], page 83 for the respective options.

6.4.2.2 Grid interpolation and smoothing

On some of the grid elements, the desired value will not be found, for example the Sky value in Section 6.4.1.1 [Finding the sky value], page 94. This can happen for lots of reasons depending on the job that was to be done on a mesh, for example when a large galaxy is present in the image, no Sky value will be found for the grid elements that lie over it. However, the Sky value should be 'guessed' over that part of the image. We cannot just ignore those regions! To fill in such blank grid elements, we use interpolation.

Parametric interpolations like bi-linear, bicubic or spline interpolations are not used because they fail terribly on the edges of the image. For example see Figure 16 in Akhlaghi and Ichikawa (2015). They are also prone to cause significant gradients in the image. So to find the interpolated value for each grid element, Gnuastro will look at a certain number of the nearest neighbors. The exact number can be specified through `--numnearest`. The median value of those grid elements will be taken as the final value for each mesh. The median is chosen because on the fainter wings of bright objects, the mean can easily become biased. If the number of meshes with a good value is less than the value given to `--numnearest`, then the program will abort and notify you. In such cases you can either decrease the value to this option or set less restrictive requirements (for example a smaller `--minmodeq`, or larger/smaller meshs) at the expense of less accurate results.

By default the process above will occur on all the grid elements, not just the ones that are blank. This is done to avoid biased results on the faint wings of bright galaxies and stars (the PSF). Because their flux penetrates into the noise very slowly, it might not be possible to completely identify that flux and ignore that mesh. Therefore, if interpolation is only done on blank pixels, such false positives can cause a bias in the vicinity of bright objects, particularly after smoothing in the next step. In order to only interpolate over blank meshs and leave the values of the successful meshs untouched, the `--interponlyblank` option can be used.

Once all the grid elements are filled, the values given to all the meshs should be smoothed. This is because the median was used in the interpolation. The median is robust in the face of outliers, but there might be strong differences from one grid element to the next. By smoothing the grid, the variation between grid element to grid element will be far less. To smooth the mesh values, Gnuastro uses an average filter. An average filter is just a convolution of the mesh grid (spatial convolution with edge correction) by a kernel with all elements having an equal weight, see Section 6.2.1.1 [Convolution process], page 66. The kernel is a square. The length of the kernel edge can be set in units of meshs through the `--smoothwidth` option (which has to be an odd number as with any kernel). If a width of 1 is given for the kernel width, then no smoothing will take place.

By default the interpolation and smoothing explained above are done independently on each channel. In some circumstances, it might be preferable to do either one of these two steps on the whole image, independent of the channels. For example when there are gradients over the image and their variation over the image is stronger than the variation caused by the channels. Through the two options `--fullinterpolation` and `--fullsmooth` you can ask for interpolation or smoothing to use all the meshs in the image, not just those in the same channel of a mesh. Note that it is still very important that no mesh contain pixels from two channels. Since the pixels have been assigned to a mesh prior to these steps (see Section 6.4.2 [Tiling an image], page 95), there is no problem in this regard.

Even after smoothing, a simple visual examination of the values given to the pixels in each mesh over the image will give a very 'boxy' or pixelated impression. To our eyes it feels that it would be much better if the values could be smoothed in sub-mesh (pixel) scales to obtain a smooth variation. An example is the results of bi-linear and bi-cubic interpolations, see Figure 16 in Akhlaghi and Ichikawa (2015) for several examples. The reasons we are not doing that level of smoothing are two fold:

- The difference in value between neighboring meshs is not statistically significant. After smoothing, the variation between the neighboring mesh values will be very small. It does visually appear to be a lot when viewing in image viewers like SAO ds9. This is because such viewers use the minimum and maximum range of all the pixel values as a reference to choose the color given to each pixel. However compared to the standard deviation of the noise in the image, the different values of each mesh are completely negligible. Please confirm this for your self on your own datasets to clearly understand.

- Such pixel-based (not mesh-based) interpolation will be very time consuming. Since the difference between the neighboring meshs is statistically negligible, it is simply not worth the time investment.

6.4.2.3 Checking grid values

Programs that use the mesh grid to calculate some value, also have checking options (that start with --check). When such options are called, the program will create a specific output file depending on what you want to check. When the operation is done on the meshes, this file shows the values assigned to each mesh grid element. It has three extensions:

1. The value that was initially calculated for each grid element. If a mesh was not successful in providing a value, a NaN value is stored for that mesh.

2. The interpolated values, see Section 6.4.2.2 [Grid interpolation and smoothing], page 98.

3. The smoothed values, see Section 6.4.2.2 [Grid interpolation and smoothing], page 98.

If more than one value is calculated for each mesh (for example the mean and its standard deviation), then each step has that many extensions. By default, these check image outputs have the same size (pixels) as the input image. So all the pixels within a mesh are given the same value. This is useful if you want to later apply these values to the image through another program for example. There is a one to one correspondence between the input image pixels and the grid showing you the mesh values for that pixel.

In other situations, the input pixel resolution might not be important for you and you just want to see the relative mesh values over the image. In such cases, you can call the --meshbasedcheck option so the check image only has one pixel for each mesh. This image will only be as big as the full mesh grid and there will be no world coordinate system. When the input images are really large, this can make a difference in both the processing of the programs and in viewing the images.

Another case when only one pixel for each mesh will be useful is when you might want to display the mesh values in a document and you don't want the volume of the document (in bytes) to get too high. For example if the mesh sizes are 30 pixels by 30 pixels, then the mesh grid created through this option will take $30 \times 30 = 900$ times less space! You can resize the standard output, but the borders between the meshs will be blurred. The best case is to call that program with this option.

6.4.2.4 Mesh grid options

The mesh grid structure defined here (see Section 6.4.2 [Tiling an image], page 95) is used by more than one program. Therefore in order to avoid repetition, all the options to do with the mesh grid (and are shared by all the programs using it) are listed here.

Some programs might define multiple meshs over the image (for example in NoiseChisel, there is a large and a small mesh for different operations, see Section 7.2 [NoiseChisel], page 112), in such cases, the options for each mesh are designated by an appropriate prefix. For example in NoiseChisel the small and large mesh sizes are specified through the --smeshsize and --lmeshsize respectively. Both these options are similar to the --meshsize option explained below, but for their respective grid. Note that the short option name might also differ. If such options exist in a program, they are listed in the 'Invoking ProgramName' section within the list of options.

-s

--meshsize

> (=INT) The size of each mesh, see Section 6.4.2 [Tiling an image], page 95. If the width of all channels are not an exact multiple of the specified size, then the last mesh on each axis will have a different size to cover the full channel.

-a

--nch1   (=INT) The number of channels along the first FITS axis (horizontal when viewed in SAO ds9). If the length of the image is not an exact multiple of this number, then the program will stop. You can use ImageCrop (Section 6.1 [ImageCrop], page 58) to trim off or add some pixels (blank pixels if added, see Section 6.1.3 [Blank pixels], page 60) to the image if it is not an exact multiple.

-b

--nch2   (=INT) The number of channels along the second FITS axis, (vertical when viewed in SAO ds9). Similar to --nch1.

-L

--lastmeshfrac

> (=FLT) Fraction of extra area on the last (rightest on the first FITS axis and highest/top on the second) mesh, to define a new (smaller) one. See Section 6.4.2 [Tiling an image], page 95.

--checkmesh

> An image with suffix _mesh.fits will be created for you to check the mesh grid, see Section 6.4.2 [Tiling an image], page 95. The input image will be the first extension, followed by an extension, where each pixel is labeled (number starting from zero) by the ID of the mesh it belongs to. If the program uses multiple mesh grids, the output will have more than two extensions. By flipping through the extensions, you can check the positioning and size of the meshs.

`-d`

`--mirrordist`

> (=FLT) The distance beyond the mirror point (in units of the error in the mirror point) to check for finding the mode in each mesh. This is part of the process to quantify the presence of signal in a mesh, see Section 6.4.2.1 [Quantifying signal in a mesh], page 97. See appendix C in Akhlaghi and Ichikawa (2015) for a complete explanation of the mode-finding algorithm. The value to this option is shown as α in that appendix.

`-Q`

`--minmodeq`

> (=FLT) The minimum acceptable quantile for the mode of each mesh. The median is on the 0.5 quantile of the image and as long as we have positive signal (all astronomically relevant observations), the mode will be less than the median. The sky value is only found on meshes where the median and mode are approximately the same, see Section 6.4.1 [Sky value], page 93.

`--interponlyblank`

> Only interpolate the blank pixels. By default, interpolation will happen on all the mesh grids, not just the blank ones. See Section 6.4.2.2 [Grid interpolation and smoothing], page 98.

`-n`

`--numnearest`

> (=INT) The number of nearest grid elements with a successful sky value to use for interpolating over blank mesh elements (those that had a significant contribution of signal), see Section 6.4.2.2 [Grid interpolation and smoothing], page 98.

`--fullinterpolation`

> Do interpolation irrespective of the channels in the image, see Section 6.4.2.2 [Grid interpolation and smoothing], page 98.

`-T`

`--smoothwidth`

> (=INT) The width of the average filter kernel used to smooth the interpolated image in units of pixels. See Section 6.4.2.2 [Grid interpolation and smoothing], page 98. If this option is given a value of 1 (one), then no smoothing will be done.

`--fullsmoothing`

> Do smoothing irrespective of the channels in the image, see Section 6.4.2.2 [Grid interpolation and smoothing], page 98.

`--meshbasedcheck`

> Store the fixed value in each mesh in one check image pixel, see Section 6.4.2.3 [Checking grid values], page 99. Note that this image has no world coordinate system.

6.4.3 Mask image

All of the programs in Gnuastro that do processing on the input data can also account for masked pixels. Particularly in raw data processing, there are usually a set of pixels in the image that should not be included in any analysis. For example saturated pixels on the centers of bright objects, or the edges of the image which received no data and were only used for bias calculation. The detectors or the very early processing that is done on raw images clearly identifies such cases and usually assigns an integer (or flag or mask value) to those pixels. Pixels that are good for processing usually have a zero value in the mask image. It goes without saying that the two images have to have the same size. Note that if the mask image has blank pixels, then they act like pixels with non-zero values and will be masked (see Section 6.1.3 [Blank pixels], page 60).

The integer values in the mask images are usually sums of powers of two. Each power of two has a specific meaning and since the sum of two different sets of powers of two are never equal, each mask value identifies a set of different properties for each masked pixels. For some analysis, some masked properties might not be a problem. Therefore a pixel that only has that property should be included. In such cases, you can use bit flags to keep some of the powers of two and remove the rest. The individual Gnuastro programs do not consider these issues. Therefore, if some masked pixels should be included in the analysis, it is best to use another tool to set the appropriate mask pixel values to zero prior to running the analysis program. We are working on such a program as part of Gnuastro. If the data type (BITPIX) of the input mask image is not an integer type, the programs will print a warning, but continue on with the analysis. This usually happens because of a mistake in specifying the file or the HDU.

The programs that accept a mask image, all share the options below. Any masked pixels will receive a NaN value (or a blank pixel, see Section 6.1.3 [Blank pixels], page 60) in the final output of those programs. Infact, another way to notify any of the Gnuastro programs to not use a certain set of pixels in a dataset is to set those pixels equal to appropriate blank pixel value for the type of the image, Section 6.1.3 [Blank pixels], page 60.

-M
--mask   (=STR) Mask image file name. If this option is not given and the --mhdu option has a different value from --hdu, then the input image name will be used. If a name is specified on the command line or in any of the configuration files, it will be used. If the program doesn't get any mask file name, it will use all the non-blank (see Section 6.1.3 [Blank pixels], page 60) pixels in the image. Therefore, specifying a mask file name in any of the configuration files is not mandatory.

-H
--mhdu   (=STR) The mask image header name or number. Similar to the --hdu option, see Section 4.1.4 [Common options], page 34. Like --mask, this option does not have to be included in the configuration file or the command-line. However, if it is present on either of them, it will be used.

6.4.4 Invoking SubtractSky

SubtractSky will find the sky value on a grid in an image and subtract it. The executable name is astsubtractsky with the following general template:

```
$ astsubtractsky [OPTION ...] Image
```

One line examples:

```
$ astsubtractsky image.fits
$ astsubtractsky --hdu=0 --mhdu=1 image.fits
$ astsubtractsky --nch1=2 --nch2=2 image.fits
$ astsubtractsky --mask=maskforimage.fits image.fits
```

The only required input to SubtractSky is the input data file that currently has to be only a 2D image. But in the future it might be useful to use it for 1D or 3D data too. Any pixels in the image with a blank value will be ignored in the analysis, see Section 6.1.3 [Blank pixels], page 60. Alternatively a mask can be specified which indicates pixels to not be used (see --mask and --mhdu). The common options to all Gnuastro programs can be seen in Section 4.1.4 [Common options], page 34 and input data formats are explained in Section 4.1.2 [Arguments], page 32.

SubtractSky uses a mesh grid to tile the image. This enables it to deal with possible gradients, see Section 6.4.2 [Tiling an image], page 95. The mesh grid options are common to all the programs using it, and are listed in Section 6.4.2.4 [Mesh grid options], page 100. In order to ignore some pixels during the analysis, you can specify a mask image, see Section 6.4.3 [Mask image], page 102 for an explanation and the relevant options. For a more accurate result, a kernel file name can be specified so the image is first convolved, see Section 6.4.2.1 [Quantifying signal in a mesh], page 97, see Section 6.2.4 [Convolution kernel], page 83 for the kernel related options. The --kernel option is not mandatory and if not specified anywhere prior to running, the original image will be used. Please run SubtractSky with the --help option to list all the recognized options, irrespective of which part of the manual they are fully explained in.

-u

--sigclipmultip

> (=FLT) The multiple of the standard devation to clip from the distribution in σ-clipping. This is necessary to remove the effect of cosmic rays, see Section 6.4.1 [Sky value], page 93 and Section 7.1.2 [Sigma clipping], page 105.

-t

--sigcliptolerance

> (=FLT) The tolerance of sigma clipping. If the fractional change in the standard deviation before and after σ-clipping is less than the value given to this option, σ-clipping will stop, see Section 7.1.2 [Sigma clipping], page 105.

--checksky

> A two extension FITS image ending with _smooth.fits will be created showing how the interpolated sky value is smoothed.

--checkskystd

> In the interpolation and sky checks above, include the sky standard devation too. By default, only the sky value is shown in all the checks. However with this option, an extension will be added showing how the standard deviation on each mesh is finally found too.

7 Image analysis

Astronomical images contain very valuable information, the tools in this section can help in extracting and quantifying that information. For example calculating image statistics, or finding the sky value or detecting objects within an image.

7.1 ImageStatistics

The distribution of pixel values in an image can give us valuable information about the image, for example if it is a positively skewed distribution, we can see that there is significant data in the image. If the distribution is roughly symmetric, we can tell that there is no significant data in the image.

On the other hand, in some measurements that we do on the image, we might need to know the certain statistical parameters of the image. For example, if we have run a detection algorithm on an image, and we want to see how accurate it was, one method is to calculate the average of the undetected pixels and see how reasonable it is (if detection is done correctly, the average of undetected pixels should be approximately equal to the background value, see Section 6.4.1 [Sky value], page 93). ImageStatistics is built for precisely such situatons.

7.1.1 Histogram and Cumulative Freqency Plot

Histograms and the cumulative frequency plots are both used to study the distribution of data. The histogram is mainly easier to understand for the untrained eye, while the cumulative frequency plot is more accurate, but needs a good level of experience for interpretation.

A histogram shows the number of data points which lie within pre-defined intervals (bins). It is used to get a general view of the distribution and its shape. The width of the bins is one of the most important parameters for a histogram. In the limiting case that the bin-widths tend to zero (and assuming there is data for each bin), then the normalized histogram would show the probability distribution function of the distribution. Normalizing a histogram means to divide the number of data points in each bin by the total number of data.

In the cumulative frequency plot of a distribution, the x axis is the sorted data values and the y axis is the index of each data in the sorted distribution. Unlike a histogram, a cumulative frequency plot does not involve intervals or bins. This makes it less prone to any sort of bias or error that a given bin-width would have on the analysis. When a larger number of the data points have roughly the same value, then the cumulative frequency plot will become steep in that vicinity. This occurs because on the x axis (data values), there is little change while on the y axis the indexs constantly increase. Normalizing a cumultaive frequency plot means to divide each index (y axis) by the total number of data points.

Unlike the histogram which has a limited number of bins, ideally the cumulative frequency plot should have one point for every data point. Even in small images (for example a 200×200) this will result in an unreasonably larger number of points to plot (40000)! So when the cumulative frequency plot of an image is stored in a text file, it is best to only store its value on a certain number of points (intervals) rather than the whole data. The number of points to use for the final plot can be specified with the --cfpnum option.

Note that the interval's lower value is considered to be part of each interval, but its larger value is not. Formally, an interval between a and b is represented by [a, b). This is true for all the intervals except the last one. The last interval is closed or [a, b].

7.1.2 Sigma clipping

Let's assume that you have pure noise (centered on zero) with a clear Gaussian distribution, see Section 8.2.1.1 [Photon counting noise], page 142. Now let's assume you add very bright objects (signal) on the image which have a very sharp boundary. By a sharp boundary, we mean that there is a clear cutoff at the place the objects finish. In other words, at their boundaries, the objects do not fade away into the noise. In such a case, when you plot the histogram (see Section 7.1.1 [Histogram and Cumulative Freqency Plot], page 104) of the distribution, the pixels relating to those objects will be clearly separte from pixels that belong to parts of the image that did not have data. In the cumulative frequency plot, you would observe a long flat region were for a certain range of data (x axis), there is no increase in the index (y axis).

In cases like the above, σ-clipping is a very useful tool to remove the effect (bias) of such forms of signal from the distribution. In astronomical applications, cosmic rays (when they collide at a near normal incidence angle) are a very good example of such signals. The tracks they leave behind in the image are perfectly immune to the blurring caused by the atmosphere and the aperture. They are also very energetic and so their borders are usually clearly separated from the surrounding noise. So σ-clipping is very useful in removing their effect on the data. See Figure 15 in Akhlaghi and Ichikawa (2015).

σ-clipping is the very simple iteration below. In each iteration, the range of input data might decrease and so when the data in the image have the conditions above, they will be removed. The criteria to stop the iteration will be discussed below.

1. Calcuate the mean, standard deviation (σ) and median (m) of a distribution.

2. Remove all points that are smaller or larger than $m \pm \alpha\sigma$.

The reason the median is used as a reference and not the mean is that the mean is too significantly affected by the presence of signal, while the median is less affected, see Section 6.4.1.1 [Finding the sky value], page 94. As you can tell from this algorithm, besides the condition above (that the signal have clear high signal to noise boundaries) σ-clipping is only useful when the signal does not cover more than half of the full data set. If they do, then the median will lie over the signal and sigma clipping might remove the pixels with no signal.

In the literature researchers use either one of two criteria to stop this iteration above:

- When a certain number of iterations has taken place.

- When the new measured standard deviation is within a certain tolerance level of the old one. The tolerance level is defined by:

$$\frac{\sigma_{old} - \sigma_{new}}{\sigma_{new}}$$

The standard deviation is heavily influenced by the presence of outliers. Therefore the fact that it stops changing between two iterations is a sign that we have successfully removed outliers. Note that in each clipping, the dispersion in the distribution is either less or equal. So $\sigma_{old} \geq \sigma_{new}$.

Other objects in astronomical data analysis like galaxies and stars are blurred by the atmosphere and the telescope aperture. Galaxies in particular do not appear to have a clear high signal to noise cutoff at all. Therefore σ-clipping will not be useful in removing their effect on the data. In astronomy, it is only useful for removing the effect of Cosmic rays.

7.1.3 Mirror distribution

The mirror distribution of a data set was defined in Appendix C of Akhlaghi and Ichikawa (2015). It is best visiualized by mentally placing a mirror on the histogram of a distribution at any point within the distribution (which we call the mirror point).

Through the `--mirrorquant` in ImageStatistics, you can check the mirror of a distribution when the mirror is placed on any given quantile. The mirror distribution is plotted along with the input distribution both as histograms and cumulative frequency plots, see Section 7.1.1 [Histogram and Cumulative Freqency Plot], page 104. Unlike the rest of the histograms and cumulative frequency plots in ImageStatistics, the text files created with the `--mirrorquant` and `--checkmode` will contain 3 columns. The first is the horizontal axis similar to all other histograms and cumulative frequency plots. The second column shows the respective value for the actual data distribution and the third shows the value for the mirror distribution.

The value for each bin of both histogram is divided by the maximum of both. For the cumulative frequency plot, the value in each bin is divided by the maximum number of elements. So one of the cumulative frequency plots reaches the maximum vertical axis of 1. The outputs will have the `_mirrorhist.txt` and `_mirrorcfp.txt` suffixs respectively. You can use a simple Python script like the one below to display the histograms and cumulative frequency plots in one plot:

```
#! /usr/bin/env python3

# Import the necessary modules:
import sys
import numpy as np
import matplotlib.pyplot as plt

# Load the two files:
a=np.loadtxt(sys.argv[1]+"_mirrorhist.txt")
b=np.loadtxt(sys.argv[1]+"_mirrorcfp.txt")

# Calculate the bin width:
w=a[1,0]-a[0,0]

# Plot the two histograms and cumulative frequency plots:
plt.bar(a[:,0], a[:,1], width=w, color="blue", linewidth=0, alpha=0.6)
plt.bar(a[:,0], a[:,2], width=w, color="green", linewidth=0, alpha=0.4)
plt.plot(b[:,0], b[:,1], linewidth=2, color="blue")
plt.plot(b[:,0], b[:,2], linewidth=2, color="green")

# Write the axis labels:
plt.ylim([0,np.amax(a[:,1])])
```

```
    plt.xlim([np.amin(a[:,0]),np.amax(a[:,0])])

    # Save the output to any name from the command line:
    plt.savefig(sys.argv[1]+"_plot.pdf")
```

The output format can be anything that Python's Matplotlib recognizes (for example png
or jpg are also acceptable). So, if your input data file name was `input.fits`, and you want
to see how its mirror distribution would look like if the mirror was placed at the 0.8 quantile
(or the value which is above 80 percent of your data) you can run the following sequence
of commands to see the combined cumulative frequency plot and histogram together in one
PDF file. Let's assume you have put the Python script above into the file `mirrorplot.py`.
The second command makes the Python script an executable file.

```
    $ ls
    input.fits mirrorplot.py
    $ chmod +x mirrorplot.py
    $ astimgstat input.fits --nohist --nocfp --mirrorquant=0.8
    $ ls
    input.fits  input_mirrorcfp.txt  input_mirrorhist.txt  mirrorplot.py
    $ ./mirrorplot.py input
```

By default, the range of the mirror distribution is set to the 0.01 quantile of the image
to 2 times the distance of the mode to that point. This is done so that the mode (which
will have a value of zero in this plot) is positioned exactly on 1/3rd point of the x axis plot.
The number of bins is the same value as the `--histnumbins` option. Alternatively, through
the option `--histrangeformirror`, the histogram properties set with the `--histmin` and
`--histmax` can be used. In these mirror plots, the mirror value is going to have the value
of zero and one of the bins is going to start at zero (if `--histrangeformirror` is called, the
given ranges will also shift accordingly).

7.1.4 Invoking ImageStatistics

ImageStatistics will print the major statistical measures of an image's pixel value distribution. The executable name is `astimgstat` with the following general template

```
    $ astimgstat [OPTION ...] InputImage.fits
```

One line examples:

```
    $ astimgstat input.fits
    $ astimgstat animage.fits --ignoremin --nohist
    $ astimgstat anotherimage.fits --mask=detectionlabels.fits --mhdu=1
```

If ImageStatistics is to do any data processing, an input image should be provided with
the recognized extensions as input data, see Section 4.1.2 [Arguments], page 32. See
Section 4.1.4 [Common options], page 34 for the list of options that are shared by all
programs. All the main statistical operations have their specific set of options. If a string is
given to the `--output` option, it is used as the base name for the generated files. Without
this option, the input image name is used as the name-base.

Some of the options are necessary and if they are not included in the configuration file,
ImageStatistics will not run, see Section 4.2 [Configuration files], page 37. However, for
some others this is not so: `--histmin`, `--histmax`, `--histquant`, `--cfpmin`, `--cfpmax`, `--
cfpquant`. These are options to do with the range of values in the histogram and cumulative

frequency plots. If no value is given for these options when ImageStatistics is about to start processing the data, then the full data range will be used. Such that the minimum image value will be set for the minimums and the maximum image value will be used for the maximum. The Mask name and HDU are also not mandatory in the configuration file.

By default, in verbose mode[1], along with a short summary of the basic data statistics, a simple ASCII histogram will also be printed. This can be useful for a very quick and general view of the distribution. An example verbose output of ImageStatistics on one of the $ make check outputs can be seen below:

```
$ astimgstat ./tests/convolve_spatial_warped_noised.fits          \
             --histquant=0.05
ImageStatistics started on AAA BBB CC DD:EE:FF GGGG
   - Input read: ./tests/convolve_spatial_warped_noised.fits (hdu: 0)
   -- Number of points                            10000
   -- Minimum                                     -38.2066
   -- Maximum                                     1268.72
   -- Sum                                         154927
   -- Mean                                        15.4927
   -- Standard deviation                          60.5407
   -- Median                                      4.82691
   -- Mode (quantile, value)                      0.4335, 2.90301
   -- Mode symmetricity and its cutoff value      0.5909, 17.2507
   -- ASCII histogram in the range: -13.912957  --  65.058487:
    |           * *
    |          *******
    |          ******** **
    |        *************
    |      *****************
    |     *******************
    |   ***********************
    |*************************
    |****************************
    |******************************************
    |********************************************************************
    |-------------------------------------------------------------


   - Sigma clipping results (Median, Mean, STD, Number):
   - 4.00 times sigma by convergence (tolerance: 0.2000):
      1: 4.826907, 15.492665, 60.540707, 10000
      2: 4.665353, 10.117773, 27.793823, 9881
      3: 4.433090, 7.458610, 18.510950, 9715
      4: 4.216213, 6.176818, 15.231371, 9575
      5: 4.088199, 5.679162, 14.183748, 9502
   - 4.00 sigma-clipping 5 times:
      1: 4.826907, 15.492665, 60.540707, 10000
```

---

[1] If the -q option is not called then all programs will operate in verbose mode, see Section 4.1.4 [Common options], page 34.

```
    2: 4.665353, 10.117773, 27.793823, 9881
    3: 4.433090, 7.458610, 18.510950, 9715
    4: 4.216213, 6.176818, 15.231371, 9575
    5: 4.088199, 5.679162, 14.183748, 9502
ImageStatistics finished in:  0.006964 (seconds)
```

-M

--mask   (=STR) The file name of a mask image. If this option is not given on the command line or in the configuration files and --mhdu is not given or is identical to --hdu, then no mask image will be used.

-H

--mhdu   (=STR) The header name of the mask extension.

-r

--ignoremin

> Ignore all data elements that have a value equal to the minimum value in the image. In practice this is like masking those pixels, their values will not be used.

-l

--lowerbin

> Set the first column of the histogram and cumulative frequency plots to the lower interval boundary. By default (without calling this option), the central interval value is used.

--onebinvalue

> (=FLT) Make sure that one bin starts with the value to this option. In practice, this will shift the bins used to find the histogram and cumulative frequency plot such that one bin's lower interval becomes this value. For example when the histogram range includes negative values, but the data doesn't. If zero is somewhere between one bin, then the viewers of the plot(s) will think negative data is also present. By setting --onebinvalue=0, you can make sure that the viewers of the histogram will not be confused.

> Note that by default, the first row of the histogram and cumulative frequency plot show the central values of each bin. So in the example above you will not see the 0.000. To see it, add the --lowerbin option to show the lower value of each bin. If you don't care about the bin positions within the specified range you can set the value to this option to a Not-a-Number (NaN) value on the command line (--onebinvalue=nan) or in the configuration files with a nan following the option name. If the value is not within the specified bin range, it will be ignored.

--noasciihist

> Do not show an ASCII plot on the command line.

--mirrorquant

> (=FLT) quantile to put the mirror. A value between 0 and 1. See Section 7.1.3 [Mirror distribution], page 106 for a complete explanation. Outputs two files with suffixes _mirrorhist.txt and _mirrorcfp.txt.

`--checkmode`

    The mode of the data is found by comparing the input data distribution with its mirror distribution. If this option is called, the mirror distribution's histogram and cumulative frequency plots will be saved in to plain text files ending with `_modehist.txt` and `_modecfp.txt`. See the explanation for Section 7.1.3 [Mirror distribution], page 106 for more details about these two files and how you can easily plot the outputs. This option only works when when ImageStatistics is in verbose mode. Since otherwise the mode is not calculated.

    To draw the plots you can use the script in Section 7.1.3 [Mirror distribution], page 106. Just change the appended suffixes in the two calls to `np.loadtxt` in the Python script.

`--histrangeformirror`

    Use `--histmin` and `--histmax` for the range of the mirror distributions (which are produced with the `--mirrorquant` and `--checkmode` options).

Histogram: The stored histogram is stored in a text file ending with `_hist.txt`.

`--nohist`   Do not calculate or save the histogram.

`--normhist`

    Make a normalized histogram, see Section 7.1.1 [Histogram and Cumulative Freqency Plot], page 104.

`--maxhistone`

    Divide all histogram bins by the number in the bin with the most data points. This is very useful if you want to plot the histogram along with a normalized cumulative frequency plot in one plot. Note that if the histogram numbers are important in showing along with the cumulative frequency plot, you can use `--maxcfpeqmaxhist`, see below.

`-n`

`--histnumbins`

    (=INT) The number of bins in the histogram. Note that in practice, this is also equivalent to the number of rows in the output text file.

`-i`

`--histmin`

    (=FLT) The minimum value to use in the histogram. If `--histquant` is given, then any value given `--histmin` or `--histmax` is ignored.

`-x`

`--histmax`

    (=FLT) The maximum value to use in the histogram. Similar to `--histmin`.

`-Q`

`--histquant`

    (=FLT) Set the range of the histogram based on the image quantile. So `--histquant=0.05` is given, all the data from the 0.05 quantile to 0.95 quantile will be used in the histogram. This is useful when there is a small number of outliers in the image. Note that if this option is given, any (possible) value given to `--histmin` or `--histmax` are ignored.

Cumulative Frequency Plot: The cumulative frequency plot will be stored in a text file ending with `_cfp.txt`. To be more realistic, the average of the indexs in each interval is used as the second column, see Section 7.1.1 [Histogram and Cumulative Freqency Plot], page 104.

`--nocfp`   Do not calculate or store the cumulative frequency plot.

`--normcfp`

> Normalize the cumulative frequency plot, see Section 7.1.1 [Histogram and Cumulative Freqency Plot], page 104.

`--maxcfpeqmaxhist`

> Set the maximum cumulative frequency plot value to the maximum value in the histogram (if it is to be created). This is a useful in plotting the histogram and cumulative frequency plots together when the histogram numbers are important.

`--cfpsimhist`

> Set the range of the cumulative frequency plot and the number of points to store to the same range as the histogram. If the two are to be plotted together, this is very useful, since the first axis (column) of the two will become identical.

`-p`
`--cfpnum`   (=INT) The number of points to store the cumulative frequency plot. They will be evenly distributed between the range of pixel values.

`-a`
`--cfpmin`   (=FLT) The minimum value to use for the cumulative pixel value range. If `--cfpquant` is given, then any value given `--cfpmin` or `--cfpmax` is ignored.

`-n`
`--cfpmax`   (=FLT) The maximum value to use for the cumulative pixel value range. Similar to `--cfpmin`.

`-U`
`--cfpquant`

> (=FLT) Similar to `--histquant` but for the cumulative frequency plot.

σ**-clipping:** The result of each iteration of σ-clipping will be printed in the terminal for both types of sigma clipping: A certain number of times and convergence of the standard deviation.

`--nosigclip`

> If this option is called, no σ-clipping will take place.

`-u`
`--sigclipmultip`

> (=FLT) The multiple of the standard deviation above which to clip. This value is demonstrated by α in Section 7.1.2 [Sigma clipping], page 105.

`-t`
`--sigcliptolerance`

> (=FLT) If the fractional difference of the standard deviation becomes less than this value, then σ-clipping will halt, see Section 7.1.2 [Sigma clipping], page 105.

`-g`
`--sigclipnum`

>   (=INT) The number of iterations for the case where the σ-clipping iteration
>   stops after a certain number of runs.

7.2 NoiseChisel

Once raw data have gone through the initial reduction process (through the programs in
Chapter 6 [Image manipulation], page 58). We are ready to derive scientific results out of
them. Unfortunately in most cases, the scientifically interesting targets are deeply drowned
in a sea of noise. NoiseChisel is Gnuastro's tool to detect signal in noise. Infact, NoiseChisel
was the motivation behind creating Gnuastro and has a full journal article[2] devoted to
its techniques. Following the explanations for the options in Section 7.2.1.1 [NoiseChisel
options], page 113 should also give you a relatively good idea of the steps. Currently the
paper does a very thorough job at explaining the concepts and methods of NoiseChisel with
abundant demonstrations for each step. However, the paper cannot undergo any futher
updates, so as the development of NoiseChisel evolves, this section will grow.

Detection is the process of separating signal from noise. In other words, after detection
is complete, one set of data elements (pixels in an image) will be distinguished as signal
and another set of the data elements will be noise. Segmentation is the process of labeling
the detected pixels into possibly multiple components (objects). For example when two
galaxies lie sufficiently close to each other to be detected as one object.

NoiseChisel was the first software to make use of a noise-based concept to detection and
segmentation. In this method, instead of emphasizing on the signal and trying to guess the
properties of the to-be-detected targets prior to detection (for example assuming that it is an
ellipse), the emphasis is put on the noise in the image and it imposes statistically negligible
requirements on the signal. The name of NoiseChisel is derived from the first thing it does
after thresholding the image: to erode it. In mathematical morphology, erosion on pixels
can be pictured as carving off boundary pixels. So what NoiseChisel does is similar to what
a wood chisel or stone chisel does. It is just not a hardware but software.

7.2.1 Invoking NoiseChisel

NoiseChisel will detect signal in noise. The executable name is `astnoisechisel` with the
following general template

```
$ astnoisechisel [OPTION ...] InputImage.fits
```

One line examples:

```
$ astnoisechisel input.fits
$ astnoisechisel --nch1=4 --lmesh=256 input.fits
$ astnoisechisel field.fits --mask=badpixels.fits --mhdu=1
```

If NoiseChisel is to do processing, an input image should be provided with the recognized
extensions as input data, see Section 4.1.2 [Arguments], page 32. The options that are shared
by all Gnuastro programs can be seen in Section 4.1.4 [Common options], page 34. In order
to ignore some pixels during the analysis, you can specify a mask image, see Section 6.4.3

---

[2] It is currently under production by the Astrophysical Journal Supplement Series. It can also be read in
arXiv: `http://arxiv.org/abs/1505.01664`.

[Mask image], page 102 for an explanation and the relevant options. A masked pixel is completely ignored.

A convolution kernel can also be optionally given. If a value (file name) is given to `--kernel` on the command line or in a configuration file (see Section 4.2 [Configuration files], page 37), then that file will be used to convolve the image prior to thresholding. Otherwise a default kernel will be used. The default kernel is a 2D Gaussian with a FWHM of 2 pixels truncated at 5 times the FWHM. See Section 3.1.1 of Akhlaghi and Ichikawa (2015) to learn why this particular kernel was chosen as default. See Section 6.2.4 [Convolution kernel], page 83 for kernel related options.

NoiseChisel uses a mesh grid to tile the image. This enables it to deal with possible gradients, see Section 6.4.2 [Tiling an image], page 95. The mesh grid options are common to all the programs using it, and are listed in Section 6.4.2.4 [Mesh grid options], page 100. In particular, NoiseChisel uses two mesh grids: a large and a small one. The sizes of the meshs in each grid can be specified with the following two options (the `--meshsize` option is not recognized by NoiseChisel). The rest of the options explained in Section 6.4.2.4 [Mesh grid options], page 100 apply to both grids.

`-s`
`--smeshsize`

> Similar to `--meshsize` in Section 6.4.2.4 [Mesh grid options], page 100, but for the smaller mesh grid, which is most dependent on the gradients in the image.

`-l`
`--lmeshsize`

> Similar to `--meshsize` in Section 6.4.2.4 [Mesh grid options], page 100, but for the larger mesh grid, used for detection and segmentation Signal to noise ratio analysis.

Please run NoiseChisel with the `--help` option to list all the recognized options with a short explanation, irrespective of which part of the Gnuastro manual they are fully explained in, see Section 4.6 [Getting help], page 43.

7.2.1.1 NoiseChisel options

The options particular to NoiseChisel are listed below. They are classified by context and also sorted in the same order that the operations are done on the image. See Akhlaghi and Ichikawa (2015) for a very complete, detailed and illustrated explanation of each step. Reading through the option explanations should be enough to optain a general idea of how NoiseChisel works. Before the procedures explained by these options begin, the image is convolved with a kernel. The first group of options listed below are those that apply to both the detection and the segmentation processes.

`-E`
`--skysubtracted`

> If this option is called, it is assumed that the image has already been sky subtracted once. Knowing if the sky has already been subtracted once or not is very important in estimating the Signal to noise ratio of the detections and clumps. In short an extra σ_{sky}^2 must be added in the error (noise or denominator in the Signal to noise ratio) for every flux value that is present in the calculation.

This can be interpreted as the error in measuring that sky value when it was subtracted by any other program. See Section 3.3 in Akhlaghi and Ichikawa (2015) for a complete explanation.

-B

--minbfrac

(=FLT) Minimum fraction (value between 0 and 1) of blank (undetected) area in a mesh for it to be considered in measuring the following properties.

- Measuring the Signal to noise ratio of false detections during the false detection removal.

- Measuring the sky value (average of undetected pixels). Both before the removal of false detections and after it.

- Clump Signal to noise ratio in the blank regions.

Because of the PSF, astronomical objects, other than cosmic rays, never have a clear cutoff and commonly sink into the noise very slowly. Even below the very low thresholds used by NoiseChisel. So when a large fraction of the area of one mesh is covered by detections, it is very probable that their faint wings are present in the undetected regions. Therefore, to get an accurate measurement of the above parameters over the full mesh grid, meshs that harbor too many detected regions should be excluded.

-F

--minnumfalse

(=INT) The minimum number of 'psudo-detections' (in identifying false detections) or clumps (in identifying false clumps) in each large mesh grid. If their number is less than this value, this mesh will be left blank and filled during mesh interpolation, see Section 6.4.2.2 [Grid interpolation and smoothing], page 98.

The Signal to noise ratio of false detections and clumps in each mesh is found using the quantile of the Signal to noise ratio distribution of the psudo-detections and clumps over the undetected pixels in each mesh. If the number of Signal to noise ratio measurements in each mesh is not enough, the quantile will not be accurate. For example if you set --detquant=0.99 (or the top 1 percent), then it is best to have at least 100 Signal to noise ratio measurements.

Detection is the process of separating the pixels in the image into two groups: 1) Signal and 2) Noise. Through the parameters below, you can customize the detection process in NoiseChisel.

-t

--qthresh

(=FLT) The quantile threshold to apply to the convolved image. The detection process begins with applying a quantile threshold to each of the small mesh grid elements, see Section 6.4.2 [Tiling an image], page 95. The quantile is only calcuated for those meshs that don't have any significant signal within them, see Section 6.4.2.1 [Quantifying signal in a mesh], page 97.

The quantile value is a floating point value between 0 and 1. Assume that we have sorted the N data elements of a distribution (the pixels in each mesh on the convolved image). The quantile (q) of this distribution is the value of the element with an index of (the nearest integer to) $q \times N$ in the sorted data set. After thresholding is complete, we will have a binary (two valued) image. The pixels above the threshold are known as foreground pixels (have a value of 1) while those which lie below the threshold are known as background (have a value of 0).

`--checkthreshold`

Check the quantile threshold values on the mesh grid. A file suffixed with `_thresh.fits` will be created, see Section 6.4.2.3 [Checking grid values], page 99.

`-e`
`--erode` (=`INT`) The number of erosions to apply to the binary thresholded image. Erosion is simply the process of flipping (from 1 to 0) any of the foreground pixels that neighbor a background pixel. In a 2D image, there are two kinds of neighbors, 4-connected and 8-connected neighbors. You can specify which type of neighbors should be used for erosion with the `--erodengb` option, see below.

Erosion has the effect of shrinking the foreground pixels. To put it another way, it expands the holes. This is a founding principle in NoiseChisel: it exploits the fact that with very low thresholds, the holes in the very low surface brightnesss regions of an image will be smaller than regions that have no signal. Therefore by expanding those holes, we are able to separate the regions harboring signal.

`--erodengb`

(=`4or8`) The type of neighborhood (structuring element) used in erosion, see `--erode` for an explanation on erosion. In 4-connectivity, the neighbors of a pixel are defined as the four pixels on the top, bottom, right and left of a pixel that share an edge with it. The 8-connected neighbors on the other hand include the 4-connected neighbors along with the other 4 pixels that share a corner with this pixel. See Figure 6 (a) and (b) in Akhlaghi and Ichikawa (2015) for a demonstration.

`-p`
`--opening`

(=`INT`) Depth of opening to be applied to the eroded binary image. Opening is a composite operation. When opening a binary image with a depth of n, n erosions (explained in `--erode`) are followed by n dilations. Simply put, dilation is the inverse of erosion. When dilating an image any background pixel is flipped (from 0 to 1) to become a foreground pixel. Dilation has the effect of fattening the foreground. Note that in NoiseChisel, the erosion which is part of opening is independent of the initial erosion that is done on the thresholded image (explained in `--erode`). The structuring element for the opening can be specified with the `--openingngb` option. Opening has the effect of removing the thin foreground connections (mostly noise) between separate foreground 'islands' (detections) thereby completely isolating them. Once opening is complete, we have *initial* detections.

`--openingngb`

(=`4or8`) The structuring element used for opening, see `--erodengb` for more information about a structuring element.

`-u`

`--sigclumpmultip`

> (=FLT) The multiple of the standard deviation during σ-clipping. NoiseChisel uses σ-clipping to remove the effect of cosmic rays when calculating the average and standard deviation of the undetected regions.
>
> Since cosmic rays have sharp boundaries and are usually small, the erosion and opening might put them within the undetected pixels. Although they might cover a very small number of pixels, they usually have very large flux values which can easily bias the average and standard devation measured on a mesh. Their effect can easily be removed by σ-clipping, see Section 7.1.2 [Sigma clipping], page 105. NoiseChisel uses the convergence of the value of the standard deviation as the criteria to stop the σ-clipping iteration.

`-r`

`--sigcliptolerance`

> (=FLT) The tolerance level to stop σ-clipping. The iteration is stopped when $(\sigma_{old} - \sigma_{new})/\sigma_{new}$ becomes smaller than the value given to this option. Note that σ_{old} will always be larger than σ_{new}. Only statistical scatter (error) can cause it to be smaller, in which case they can be considered to be approximately equal.

`--checkdetectionsky`

> Check the initial approximation of the sky value and its standard deviation in a FITS file ending with **_detsky.fits**. See Section 6.4.2.3 [Checking grid values], page 99 for more information. If **--meshbasedcheck** is not called, then the first extension will be the the binary image with initial detections labeled one and background labeled zero. The mesh values will be in the subsequent extensions.

`-R`

`--dthresh`

> (=FLT) The detection threshold: a multiple of the initial sky standard deviation added with the initial sky approximation. This flux threshold is applied to the initially undetected regions on the unconvolved image. The background pixels that are completely engulfed in a 4-connected foreground region are converted to background (holes are filled) and one opening (depth of 1) is applied over both the initially detected and undetected regions. The Signal to noise ratio of the resulting 'psudo-detections' are used to identify true vs. false detections. See Section 3.1.5 and Figure 7 in Akhlaghi and Ichikawa (2015) for a very complete explanation.

`-i`

`--detsnminarea`

> (=INT) The minimum area to calculate the Signal to noise ratio on the psudo-detections of both the initially detected and undetected regions. When the area in a psudo-detection is too small, the Signal to noise ratio measurements will not be accurate and their distribution will be heavily skewed to the postive. So it is best to ignore any psudo-detection that is smaller than this area.

`--detsnhistnbins`

>   (=INT) If not equal to zero, a histogram of the Signal to noise ratios of the psudo-detections will be stored in a text file for every large mesh grid element which is successful in defining a Signal to noise ratio in the grid. The number of bins in this histogram is specified by the value given to this option. This is good for inspecting the best possible value to `--detsnminarea`. The best values are obtained when the distribution is not skewed significantly. Note that when correlated noise is present (processed images) a certain level of skewness will be present in any case.

`-c`
`--detquant`

>   (=FLT) The quantile of the Signal to noise ratio distribution of the psudo-detections in each mesh to use for filling the large mesh grid. Note that this is only calculated for the large mesh grids that satisfy the minimum fraction of undetected pixels (value of `--minbfrac`) and minimum number of psudo-detections (value of `--minnumfalse`).

`--checkdetectionsn`

>   Check the Signal to noise ratio value on each large mesh grid in a file ending with `_detsn.fits`, see Section 6.4.2.3 [Checking grid values], page 99 for more information.

`-I`
`--dilate`   (=INT) Number of times to dilate the final true detections. See the explanations in `--opening` for more information on dilation. The structuring element for this final dilation is fixed to an 8-connected neighborhood. This is because astronomical objects, except cosmic rays, never have a clear cutoff, so all the 8-pixels connected to the border pixels of a detection might harbor data.

`--checkdetection`

>   Every step of the detection process will be added as an extension to a file with the suffix `_det.fits`. Going through each would just be a repeat of the explanations above and also of those in Akhlaghi and Ichikawa (2015). The extension label should be sufficient to recognize which step you are observing. Viewing all the steps can be the best guide in choosing the best set of parameters. Note that calling this function will significantly slow NoiseChisel.

`--checksky`

>   Check the final sky and its standard deviation values on the mesh grid. Similar to `--checkdetectionsky`.

`--saveskysubed`

>   Save the sky subtracted image (where the sky was calculated from the average of undetected pixels, see Section 6.4.1 [Sky value], page 93) into a file ending with `_skysubed.fits`.

`--checkmaskdet`

>   Mask (set to NaN) the undetected pixels in one extension and the detected pixels in the next extension of a file ending with `_maskdet.fits`.

Segmentation is the process of possibly breaking up a detection into multiple objects and clumps. In deep surveys segmentation becomes particularly important since galaxies might fall along the same line of sight or they might be merging. It is thus very important to be able separate the pixels within a detection if it is necessary. After segmentation, such a detected region will get different labels.

In NoiseChisel, segmentation is done by first finding the 'true' clumps over a detection and then expanding those clumps to a certain flux limit. True clumps are found in a process very similar to the true detections explained above, see Akhlaghi and Ichikawa (2015) for more information. If the connections between the grown clumps are weaker than a given threshold, the grown clumps are considered to be separate objects. Otherwise, the are considered to be part of the same object.

`--detectonly`

> If this option is called, no segmentation will be done. The object labels extension in the output will simply be the detection (connected components) labels and the clumps image will be blank (see Section 7.2.1.2 [NoiseChisel output], page 120).
>
> This option can result in faster processing when only the noise properties of the image are desired for a catalog using another image's labels. A common case is when you want to measure colors or SEDs in several images. Let's say you have images in two colors: A and B. For simplicity also assume that they are exactly on the same position in the sky with the same pixel scale.
>
> You choose to set A as a reference, so you run the full NoiseChisel on A. Then you run NoiseChisel on B with this option since you only need the noise properties of B (for the signal to noise column in its catalog). You can then run MakeCatalog on A normally, see Section 7.3 [MakeCatalog], page 121. To run MakeCatalog on B, you simply set the object and clump labels images to those that NoiseChisel produced for A, see Section 7.3.2 [Invoking MakeCatalog], page 123.

`-m`

`--segsnminarea`

> (=INT) The minimum area which a clump in the undetected regions should have in order to be considered in the clump Signal to noise ratio measurement. If this size is set to a small value, the Signal to noise ratio of false clumps will not be accurately found. Note that this has to be larger than the value to `--detsnminarea`. Because the clumps are found on the convolved (smoothed) image while the psudo-detections are found on the input image.

`--checkclumpsn`

> Check the limiting clump Signal to noise ratio for true detections in a file ending with `_clumpsn.fits`, see Section 6.4.2.3 [Checking grid values], page 99 for more information on checking values on the mesh grid.

`-g`

`--segquant`

> (=FLT) The quantile of the noise clump Signal to noise ratio distribution. This value is used to identify true clumps over the detected regions.

`--segsnhistnbins`

> (=INT) Similar to `--detsnhistnbins`, but for the distribution of the Signal to noise ratio of clumps over the undetected regions.

`-v`

`--keepmaxnearriver`

> Keep a clump whose maximum flux is 8-connected to a river pixel. By default such clumps over detections are considered to be noise and are removed irrespective of their brightness (see Section 8.1.3 [Flux Brightness and magnitude], page 134). Over large profiles, that sink into the noise very slowly, noise can cause part of the profile (which was flat without noise) to become a very large and with a very high Signal to noise ratio. In such cases, the pixel with the maximum flux in the clump will be immediately touching a river pixel.

`-G`

`--gthresh`

> (=FLT) Threshold (multiple of the sky standard deviation added with the sky) to stop growing true clumps. Once true clumps are found, they are set as the basis to segment the detected region. They are grown until the threshold specified by this option.

`--grownclumps`

> In the output (see Section 7.2.1.2 [NoiseChisel output], page 120) store the grown clumps (or full detected region if only one clump was present in that detection). By default the original clumps are stored as the third extension of the output, if this option is called, it is replaced with the grown clump labels.

`-y`

`--minriverlength`

> (=INT) The minimum length of a river between two grown clumps for it to be considered in Signal to noise ratio estimations. Similar to `--segsnminarea` and `--detsnminarea`, if the length of the river is too short, the Signal to noise ratio can be noisy and unreliable. Any existing rivers shorter than this length will be considered as non-existant, independent of their Signal to noise ratio. Since the clumps are grown on the input image, this value should best be similar to the value of `--detsnminarea`. Recall that the clumps were defined on the convolved image so `--segsnminarea` was larger than `--detsnminarea`.

`-O`

`--objbordersn`

> (=FLT) The maximum Signal to noise ratio of the rivers between two grown clumps in order to consider them as separate 'objects'. If the Signal to noise ratio of the river between two grown clumps is larger than this value, they are defined to be part of one 'object'. Note that the physical reality of these 'objects' can never be established with one image, or even multiple images from one broad-band filter. Any method we devise to define 'object's over a detected region is ultimately subjective.
>
> Two very distant galaxies or satellites in one halo might lie in the same line of sight and be detected as clumps on one detection. On the other hand, the connection (through a spiral arm or tidal tail for example) between two parts

of one galaxy might have such a low surface brightness that they are broken up into multiple detections or objects. Infact if you have noticed, exactly for this purpose, this is the only Signal to noise ratio that the user gives into NoiseChisel. The 'true' detections and clumps can be objectively identified from the noise characteristics of the image, so you don't have to give any hand input Signal to noise ratio.

`--checksegmentation`

A file with the suffix `_seg.fits` will be created. This file keeps all the relevant steps in finding true clumps and segmenting the detections in various extensions. Having read the paper or the steps above, the extension name should be enough to understand which step each extension is showing. Examing this file can be an excellent guide in choosing the best set of parameters. Note that calling this function will significantly slow NoiseChisel.

7.2.1.2 NoiseChisel output

The default name and directory of the outputs are explained in Section 4.5 [Automatic output], page 43. NoiseChisel's default output (when none of the options starting with `--check` or the `--output` option are called) is one file ending with `_labeled.fits`. This file has the extensions listed below:

1. A copy of the input image, a copy is placed here for the following reasons:

   - By flipping through the extensions, a user can check how accurate the detection and segmentation process was.

   - All the inputs to MakeCatalog (see Section 7.3 [MakeCatalog], page 121) are included in this one file which makes the running of MakeCatalog after NoiseChisel very easy.

   - All masked pixels given to NoiseChisel will have a value of NaN in this image. So if you decide to run MakeCatalog after NoiseChisel, there is no more need to feed the mask image to MakeCatalog too.

2. The object labels. Each pixel in the input image is given a label in this extension, the labels start from one. The total number of labels is stored as the value to the `NOBJS` keyword in the header of this extension. It is also printed in verbose mode.

3. The clump labels. All the pixels in the input image that belong to a true clump are given a positive label in this extension. The detected regions that were not a clump are given a negative value to clearly identify the noise from the detections. The total number of clumps in this image is stored in the `NCLUMPS` keyword of this extension and printed in verbose output.

   If the `--grownclumps` option is called, or a value of `1` is given to it in any of the configuration files, then instead of the original clump regions, the grown clumps will be stored in this extension. Note that if there is only one clump (or no clumps) over a detected region, then the whole detected region is given a label of 1.

4. The final sky value on each pixel. See Section 6.4.1 [Sky value], page 93 for a complete explanation. See Section 6.4.2.2 [Grid interpolation and smoothing], page 98 for an explanation on the boxy appearance of this image.

5. Similar to the previous mesh but for the standard deviation on each pixel.

7.3 MakeCatalog

Detecting and segmenting signal over an image results in labeled images where each pixel is labeled with the ID (an integer) that is specified by the detector program. But this labeled image by its self can hardly be of any scientific use. The job of MakeCatalog is to combine the input image, the noise properties and the labels of pixels into a catalog (a text file table) which can easily be used for high-level scientific interpretations.

NoiseChisel (Gnuastro's signal detection tool, see Section 7.2 [NoiseChisel], page 112) does not produce any catalog of the detected objects by its self. Only a labeled FITS image is output, see Section 7.2.1.2 [NoiseChisel output], page 120. The output of NoiseChisel can be directly fed into MakeCatalog to generate the catalog. Some of the reasons for making the catalog in a separate program[3] are listed below:

- Complexity: Adding in a catalog functionality to the detector program will add several more steps (and options) to its processings that can equally well be done outside of it. This makes following the code harder for a curious reader and also potentially adds bugs.

  Another advantage of less complexity is when the parameter you want to measure over one profile is not provided by the developers of MakeCatalog. You can simply open this tiny little program and add your desired calculation easily. However, if making a catalog was part of NoiseChisel, it would require a lot of energy to understand all the steps in order to add desired parameter.

- Low level nature of Gnuastro: Making catalogs is a separate process from labeling (detecting and segmenting) the pixels. A user might want to do certain operations on the labed regions before creating a catalog for them. Another user might want the properties of the same pixels in another image (possibly from another broadband filter) for measuring the colors or SEDs for example.

  Here is an example of doing both: suppose you have images in various broad band filters at various resolutions and orientations. The image of one color will thus not lie exactly on another or even be in the same scale. However, it is imperative that the same pixels be used in measuring the colors of galaxies.

  Therefore NoiseChisel can be run on the reference image and ImageWarp (Section 6.3 [ImageWarp], page 86) can be applied to the labeled images to find the pixels to use in the other image. Then MakeCatalog can generate the final catalog for both targets. It is currently customary to warp the images to the same pixel grid, however, this is very harmful for the data and creates correlated noise. It is much more accurate to do the transformations on the labeled image.

7.3.1 Depth and limiting magnitude

Different observations have different noise properties and different detection methods (or one method with a different set of parameters) will have different abilities to detect certain objects in an image. Therefore it is very important that there be a scale on which we can compare different observations (images) and detection methods to objectively quantify the noise and our ability to detect signal in it.

---

[3] Most existing software that do object detection also output a catalog, so this is not a common practice.

Due to the presence of correlated noise in processed images (which are used for scientific deductions), we cannot simply deduce the limiting signal properties from those of the noise. Hence a different measure is needed for each. To quantify the level of noise, we define *depth* and to quantify the ability to reliably detect/study objects with that methodology we define the magnitude limit. In astronomy, it is common to use the magnitude (a unit-less scale) and not physical units, see Section 8.1.3 [Flux Brightness and magnitude], page 134. Therefore both these measures will be in units of magnitudes, but since magnitudes don't have units, we are just showing them like units as a place holder for clarity.

Depth [magnitude] As we make more observations on one region of the sky and add the images over each other, we are able to decrease the standard deviation of the noise in each pixel[4]. Qualitatively, this decrease manifests its self by making fainter (per pixel) parts of the objects in the image more visible. Quantitatively, it increases the Signal to noise ratio, since the signal is fixed but the noise is decreased. It is very important to have in mind that noise is defined per pixel (or independent data measurement unit), not per object.

You can think of the noise as muddy water that is completely covering a flat ground[5] with some hills[6] in it. Let's assume that in your first observation the muddy water has just been stirred and you can't see anything through it. As you wait and make more observations, the mud settles down and the *depth* of the transparent water increases as you wait. The summits of hills begin to appear. As the depth of clear water increases, the parts of the hills with lower hights can be seen more clearly.

The outputs of NoiseChisel include the Sky standard deviation (σ) on every group of pixels (a mesh) that were calculated from the undetected pixels in that mesh, see Section 6.4.2 [Tiling an image], page 95 and Section 7.2.1.2 [NoiseChisel output], page 120. Let's take σ_m as the median σ over the successful meshes in the image (prior to interpolation or smoothing, see Section 6.4.2.2 [Grid interpolation and smoothing], page 98). Note that even though on different instruments, pixels have different physical sizes (for example in μm), nevertheless, a pixel is the unit of data collection. Therefore, as far as noise is considered, the physical or projected size of the pixels is irrelevant. We thus define the *depth* of each dataset as the magnitude of σ_m.

As an example, the XDF survey covers part of the sky that the Hubble space telescope has observed the most (for 85 orbits) and is consequently very small (~ 4 arcmin^2). On the other hand, the CANDELS survey, is one of the widest multi-color surveys covering several fields (about 720 arcmin^2) but its deepest fields have only 9 orbits observation. The depth of the XDF and CANDELS-deep surveys in the near infrared WFC3/F160W filter are respectively 34.40 and 32.45 magnitudes. In a single orbit image, this same field has a depth of 31.32. Note that a larger magnitude corresponds to less brightness, see Section 8.1.3 [Flux Brightness and magnitude], page 134.

---

[4] This is true for any noisy data, not just astronomical images

[5] The ground is the sky value in this analogy, see Section 6.4.1 [Sky value], page 93. Note that this analogy only holds for a flat sky value across the surface of the image or ground.

[6] The hills are astronomical objects in this analogy. Brighter parts of the object are higher from the ground.

Magnitude limit

> [magnitude] Because of noise and methodology, no detection algorithm can be perfect and this parameter specifies the limits of the combined data and methodology used for detection. Assuming a fixed shape for all the targets, for any algorithm and its accompanying set of input parameters, there is always a certain magnitude limit below which the detections (pseudo-detections or clumps in NoiseChisel, see Section 7.2 [NoiseChisel], page 112) are no longer usable/reliable.
>
> While adding more data sets does have the advantage of decreasing the standard deviation of the noise, it also produces correlated noise. Correlated noise is produced because the raw data sets are warped (rotated, shifted or resamapled in general, see Section 6.3 [ImageWarp], page 86) before they are added with each other. This correlated noise manifests as a 'smoothing' or 'blurring' over the image. Therefore pixels in added images are no longer separate or independently measured data elements, they are 'correlated' and this produces a hurdle in our ability to detect objects in them.
>
> To find the limiting magnitude, you have to use the output of MakeCatalog and plot the number of objects as a function of magnitude with your favoriate plotting tool, this is called a "number count" plot. It is simply a histogram of the catalog in each magnitude bin. This histogram can be used in many ways to specify a magnitude limit, for example see Akhlaghi et al. (2015, in preparation) for one method of using multiple depth images in order to find this limit.

For any data-set and detection algorithm (with a specific set of parameters), the depth and limiting magnitudes can differ. The first is reported in the comments section of the catalog plain text file. Note that the accuracy in estimating the zero-point magnitude is a very important factor in an accurate comparison between magnitudes measured for different images, on possibly different instruments and cameras, see Section 8.1.3 [Flux Brightness and magnitude], page 134.

7.3.2 Invoking MakeCatalog

MakeCatalog will make a catalog from an input image and a labeled image. The executable name is `astmkcatalog` with the following general template

```
$ astmkcatalog [OPTION ...] InputImage.fits
```

One line examples:

```
$ astmkcatalog -mdri input.fits
$ astmkcatalog --floatprecision=5 input.fits
$ astmkcatalog --output=cat input_labeled.fits
$ astmkcatalog --objlabs=K_labeled.fits --objhdu=1 \
               --clumplabs=K_labeled.fits --clumphdu=2 i_band.fits
```

If MakeCatalog is to do processing, an input image should be provided with the recognized extensions as input data, see Section 4.1.2 [Arguments], page 32. Optionally a mask file can be specified to ignore some of the pixels in the image, see Section 6.4.3 [Mask image], page 102. Note that if you generated the labeled image using NoiseChisel with a mask image as input, there is no more need to inform MakeCatalog of the mask image, see Section 7.2.1.2

[NoiseChisel output], page 120. The options common to all Gnuastro programs are explained in Section 4.1.4 [Common options], page 34.

By default two catalogs will be made: one for the objects (suffixed with _o.txt) and another for the clumps (suffixed with _c.txt). Therefore if any value is given to the --output option, MakeCatalogs will simply append the two suffixes to it as the output file names. So if you want to specify an output name, it is best that it not have any suffix. If no value is given to the --output option, MakeCatalog will use the input name, see Section 4.5 [Automatic output], page 43.

The first set of options specify the properties of the inputs. Other necessary input images are treated very much like a mask image, see Section 6.4.3 [Mask image], page 102. If no name is specified for them, their HDU is checked and if that differs from the input HDU, then there is no need to specify a file name for them. The object and column label images or segmentation maps should not be of a floating point type (BITPIX).

-O

--objlabs

(=STR) The file name of the object labels, if in the same file as input, it is not mandatory.

--objhdu   (=STR) The HDU of the object labels image, the header keyword NOBJS must be present in this extension. The value to this keyword is used as the final number of objects and the number of rows in the output objects catalog. Only pixels with values above zero will be considered.

-c

--clumplabs

(=STR) Similar to --objlabs but for the labels of the clumps.

--clumphdu

(=STR) Similar to --objhdu, but for the clumps. The NCLUMPS keyword in this header specifies the number of recognized clumps.

-s

--skyfilename

(=STR) File name of an image keeping the Sky value for each pixel.

--skyhdu   (=STR) The HDU of the Sky value image.

-t

--stdfilename

(=STR) File name of image keeping the Sky value standard deviation for each pixel.

--stdhdu   (=STR) The HDU of the Sky value standard deviation image.

-z

--zeropoint

(=FLT) The zero point magnitude for the input image, see Section 8.1.3 [Flux Brightness and magnitude], page 134.

```
-E
--skysubtracted
```
> If the image has already been sky subtracted by another program, then you need to notify MakeCatalog through this option. Note that this is only relevant when the Signal to noise ratio is to be calculated.

Through the next group of options, you can customize the general the output plain text catalog. The basic idea behind the options about the width and precision of the different types is the fact that some columns don't need too much space, while some do. The width is the number of text columns given to data of each type in the output plain text catalog.

The precision is the number of digits to show after the decimal point in floating point numbers. We have defined two types of floating point numbers here, one is for less accurate precision, like magnitude, while the other is used when more accuracy is necessary. A common example of the latter is right ascension and declination. These variables usually need to be printed with more than 6 point accuracy. Note that all the values are calculated and stored internally as double precision floating point numbers, the distinction made here is only for printing them.

```
--nsigmag
```
> (=FLT) The magnitude of the value to this option multiplied by the maximum standard deviation over the objects or clumps is reported in the output catalog. This value is a per-pixel value, not per object and is not found over an area or aperture, like the common 5σ values that are commonly reported as a measure of depth. They are based on a certain area and are relics from the time of analog data collection and processing. Modern tools use digital imaging detectors and the area is that of a pixel.

```
--intwidth
```
> (=INT) The width of printing the integer values. In MakeCatalog, all IDs, numbers (counters) and areas are considered to be an integer.

```
--floatwidth
```
> (=INT) The width of a normal precision floating point column. Any column that is not designated in `--intwidth` or `--accuwidth` is considered to be a normal precision floating point.

```
--accuwidth
```
> (=INT) The width columns to be printed with extra accuracy. In MakeCatalog the following columns are printed with extra accuracy: right ascensions, declinations, brightnesses, river pixel averages (see Akhlaghi and Ichikawa 2015 for the definition of river pixels), the sky and the sky standard deviation.

```
--floatprecision
```
> (=INT) The number of digits to the right of the decimal point in normal floating point display.

```
--accuprecision
```
> (=INT) The number of digits to the right of the decimal point in more accurate floating point display.

The final group of options particular to MakeCatalog are those that specfy which columns should be displayed in the output catalogs. For each column there is an option, if it has been called on the command line or in any of the configuration files, it will included as a column in the output catalog. Some of the options apply to both objects and clumps and some are particular to only one of them. The latter cases are explicitly marked with [Objects] or [Clumps] to specify the catalog they will be placed in.

The order of the columns in the output catalog is the inverse of the order their options are read in. For example see the following command[7]:

```
$ astmkcatalog --magnitude --dec --ra --id input.fits
```

If no columns are specified in any of the configuration files (see Section 4.2 [Configuration files], page 37), then the columns in the output catalogs will have the following order: ID, RA, Dec and Magnitude. Contrary to what it looks like, this is done to make life easier for the users. The configuration files can also keep any of the columns (so you don't have to specify your desired columns every time). This inverse ordering thus comes from their precedence, see Section 4.2.2 [Configuration file precedence], page 38.

For example catalogs usually have atleast an ID column and position columns (in the image and/or the world coordinate system). By reading the order of the columns in reverse you can have your fixed set of columns in your system wide configuration file and in any particular run, if you want some other information about objects or clumps, you can add those columns on the command line. Through the user and current directory configuration files, you can also have custom catalogs in each of your working directories, without bothering to specify the columns every time.

`--i`

`--id`   The ID of the clump or object.

`-j`

`hostobjid`

          [Clumps] The ID of the object which hosts this clump.

`-I`

`--idinhostobj`

          [Clumps] The ID of this clump in its host object.

`-C`

`--numclumps`

          [Objects] The number of clumps in this object.

`-a`

`--area`   The area (number of pixels) in any clump or object.

`--clumpsarea`

          [Objects] The total area of all the clumps in this object.

`-x`

`--x`   The flux weighted center of all objects and clumps along the first FITS axis (horizontal when viewed in SAO ds9).

`-y`

`--y`   The flux weighted center of all objects and clumps along the second FITS axis (vertical when viewed in SAO ds9).

---

[7] This command is practically identical to the first command in the one-line examples, see Section 4.1.3 [Options], page 32 for an explanation of the concatenation of the on/off options.

`--clumpsx`

        [Objects] The flux weighted center of all the clumps in this object along the first FITS axis.

`--clumpsy`

        [Objects] The flux weighted center of all the clumps in this object along the second FITS axis.

`-r`

`--ra`   Right ascension of all objects or clumps.

`-d`

`--dec`   Declination of all objects or clumps.

`--clumpsra`

        [Objects] Right ascension of all clumps in this object.

`--clumpsdec`

        [Objects] Declination of all clumps in this object.

`-b`

`--brightness`

        The brightness (sum of all pixel values), see Section 8.1.3 [Flux Brightness and magnitude], page 134. For each clump in the clumps catalog, the ambient brightness (flux of river pixels around the clump multiplied by the area of the clump) is removed, see `--riverflux`. So the sum of clump brightnesses in the clump catalog will be smaller than the total clump brightness in the `--clumpbrightness` column of the objects catalog.

`--clumpbrightness`

        [Objects] The total brightness of the clumps within an object. This is simply the sum of the pixels associated with clumps in the object.

`-m`

`--magnitude`

        The magnitude of all clumps or objects, see `--brightness`.

`--clumpsmagnitude`

        [Objects] The magnitude of all clumps in this object, see `--clumpbrightness`.

`--riverave`

        [Clumps] The average flux of the river pixels of this clump. River pixels were defined in Akhlaghi and Ichikawa 2015. In short they are right on the boundaries of the clumps. This value is used internally to find the signal to noise ratio of the clumps and can also be used as a scale to guage the base (ambient) flux of the clump.

`--rivernum`

        [Clumps] The number of river pixels around this clump, see `--riverflux`.

`-n`

`--sn`   The Signal to noise ratio (S/N) of all clumps or objects. See Akhlaghi and Ichikawa (2015) for the exact equations used.

--sky   The sky flux (per pixel) value under this object or clump. This is actually the mean value of all the pixels in the sky image that lie on the same position as the object or clump.

--std   The sky value standard deviation (per pixel) for this clump or object. Like --sky, this is the average of the values in the input sky standard deviation image pixels that lie over this object.

8 Modeling and fitting

In order to fully understand observations after initial analysis on the image, it is very important to compare them with the existing models to be able to further understand both the models and the data. The tools in this chapter create model galaxies and will provide 2D fittings to be able to understand the detections.

8.1 MakeProfiles

MakeProfiles will create mock astronomical profiles from a catalog, either individually or together in one output image. In data analysis, making a mock image can act like a calibration tool, through which you can test how successfully your detection technique is able to detect a known set of objects. There are commonly two aspects to detecting: the detection of the fainter parts of bright objects (which in the case of galaxies fade into the noise very slowly) or the complete detection of an over-all faint object. Making mock galaxies is the most accurate (and idealistic) way these two aspects of a detection algorithm can be tested. You also need mock profiles in fitting known functional profiles with observations.

MakeProfiles was initially built for extra galactic studies, so currently the only astronomical objects it can produce are stars and galaxies. We welcome the simulation of any other astronomical object. The general outline of the steps that MakeProfiles takes are the following:

1. Build the full profile out to its truncation radius in a possibly over-sampled array.

2. Multiply all the elements by a fixed constant so its total magnitude equals the desired total magnitude.

3. If --individual is called, save the array for each profile to a FITS file.

4. If --nomerged is not called, add the overlapping pixels of all the created profiles to the output image and abort.

Using input values, MakeProfiles adds the World Coordinate System (WCS) headers of the FITS standard to all its outputs (except PSF images!). For a simple test on a set of mock galaxies in one image, there is no need for the third step or the WCS information.

However in complicated simulations like weak lensing simulations, where each galaxy undergoes various types of individual transformations based on their position, those transformations can be applied to the different individual images with other programs. After all the transformations are applied, using the WCS information in each individual profile image, they can be merged into one output image for convolution and adding noise.

8.1.1 Modeling basics

In the subsections below, first a review of some very basic information and concepts behind modeling a real astronomical image is given. You can skip this subsection if you are already sufficiently familiar with these concepts.

8.1.1.1 Defining an ellipse

The PSF, see Section 8.1.1.2 [Point Spread function], page 130, and galaxy radial profiles are generally defined on an ellipse so in this section first defining an ellipse on a pixelated 2D surface is discussed. Labeling the major axis of an ellipse a, and its minor axis with b, the axis ratio is defined as: $q \equiv b/a$. The major axis of an ellipse can be aligned in any direction, therefore define the angle of the major axis to the horizontal axis of the image is defined to be the position angle of the ellipse and in this manual, we show it with θ.

Our aim is to put a radial profile of any functional form $f(r)$ over an ellipse. Let's define the radial distance r_{el} as the distance on the major axis to the center of the ellipse which is located at x_c and y_c. We want to find the elliptical distance of a point located at (i, j), in the image coordinate system, from the center of the ellipse. First the coordinate system is rotated by θ to get the new rotated coordinates of that point (i_r, j_r):

$$i_r(i, j) = (i_c - i)\cos(\theta) + (j_c - j)\sin(\theta)$$

$$j_r(i, j) = (j_c - j)\cos(\theta) - (i_c - i)\sin(\theta)$$

The elliptical distance of a point located at (i, j) can now be defined as: $r_{el}^2 = \sqrt{i_r^2 + j_r^2/q^2}$. To place the radial profiles explained below over an ellipse, $f(r_{el}(i, j))$ is calculated based on the functional radial profile desired.

The way MakeProfiles builds the profile is that the nearest pixel in the image to the given profile center is found and the profile value is calculated for it, see Section 8.1.1.5 [Sampling from a function], page 133. The next pixel which the profile value is calculated on is the next nearest neighbor of the initial pixel to the profile center (as defined by r_{el}). This is done fairly efficiently using a breadth first parsing strategy[1] which is implemented through an ordered linked list.

Using this approach, we only go over one layer of pixels on the circumference of the profile to build the profile. Not one more extra pixel has to be checked. Another consequence of this strategy is that extending MakeProfiles to three dimensions becomes very simple: only the neighbors of each pixel have to be changed. Everything else after that (when the pixel index and its radial profile have entered the linked list) is the same, no matter the number of dimensions we are dealing with.

8.1.1.2 Point Spread function

Assume we have a 'point' source, or a source that is far smaller than the maxium resolution (a pixel). When we take an image of it, it will 'spread' over an area. To quantify that spread, we can define a 'function'. This is how the point spread function or the PSF of an image is defined. This 'spread' can have various causes, for example in ground based astronomy, due to the atmosphere. In practice we can never surpass the 'spread' due to the diffraction of the lens aperture. Various other effects can also be quantified through a PSF. For example, the simple fact that we are sampling in a discrete space, namely the pixels, also produces a very small 'spread' in the image.

Convolution is the mathematical process by which we can apply a 'spread' to an image, or in other words blur the image, see Section 6.2.1.1 [Convolution process], page 66. The Brightness of an object should remain unchanged after convolution, see Section 8.1.3 [Flux Brightness and magnitude], page 134. Therefore, it is important that the sum of all the pixels of the PSF be unity. The image also has to have an odd number of pixels on its sides

---

[1] http://en.wikipedia.org/wiki/Breadth-first_search

so one pixel can be defined as the center. In MakeProfiles, the PSF can be set by the two
methods explained below.

Parametric functions

A known mathematical function is used to make the PSF. In this case, only the
parameters to define the functions are necessary and MakeProfiles will make a
PSF based on the given parameters for each function. In both cases, the center
of the profile has to be exactly in the middle of the central pixel of the PSF
(which is automatically done by MakeProfiles). When talking about the PSF,
usually, the full width at half maximum or FWHM is used as a scale of the
width of the PSF.

Gaussian   In the older papers, and to a lesser extent even today, some researchers use the 2D Gaussian function to approximate the PSF of
ground based images. In its most general form, a Gaussian function
can be written as:

$$f(r) = a \exp\left(\frac{-(x-\mu)^2}{2\sigma^2}\right) + d$$

Since the center of the profile is pre-defined, μ and d are constrained. a can also be found because the function has to be normalized. So the only important parameter for MakeProfiles is the σ.
In the Gaussian function we have this relation between the FWHM
and σ:

$$\mathrm{FWHM_g} = 2\sqrt{2\ln 2}\,\sigma \approx 2.35482\sigma$$

Moffat   The Gaussian profile is much sharper than the images taken from
stars on photographic plates or CCDs. Therefore in 1969, Moffat
proposed this functional form for the image of stars:

$$f(r) = a \left[1 + \left(\frac{r}{\alpha}\right)^2\right]^{-\beta}$$

Again, a is constrained by the normalization, therefore two parameters define the shape of the Moffat function: α and β. The radial
parameter is α which is related to the FWHM by

$$\mathrm{FWHM_m} = 2\alpha\sqrt{2^{1/\beta} - 1}$$

Comparing with the PSF predicted from atmospheric turbulence
theory with a Moffat function, Trujillo et al.[2] claim that β should

[2] Trujillo, I., J. A. L. Aguerri, J. Cepa, and C. M. Gutierrez (2001). "The effects of seeing on Sérsic profiles
- II. The Moffat PSF". In: MNRAS 328, pp. 977—985.

be 4.765. They also show how the Moffat PSF contains the Gaussian PSF as a limiting case when $\beta \to \infty$.

An input FITS image

An input image file can also be specified to be used as a PSF. If the sum of its pixels are not equal to 1, the pixels will be multiplied by a fraction so the sum does become 1.

While the Gaussian is only dependent on the FWHM, the Moffat function is also dependent on β. Comparing these two functions with a fixed FWHM gives the following results:

- Within the FWHM, the functions don't have significant differences.
- For a fixed FWHM, as β increases, the Moffat function becomes sharper.
- The Gaussian function is much sharper than the Moffat functions, even when β is large.

8.1.1.3 Stars

In MakeProfiles, stars are generally considered to be a point source. This is usually the case for extra galactic studies, were nearby stars are also in the field. Since a star is only a point source, we assume that it only fills one pixel prior to convolution. In fact, exactly for this reason, in astronomical images the light profiles of stars are one of the best methods to understand the shape of the PSF and a very large fraction of scientific research is preformed by assuming the shapes of stars to be the PSF of the image.

8.1.1.4 Galaxies

Today, most practitioners agree that the flux of galaxies can be modeled with one or a few generalized de Vaucouleur's (or Sérsic) profiles.

$$I(r) = I_e exp\left(-b_n\left[\left(\frac{r}{r_e}\right)^{1/n} - 1\right]\right)$$

Gérard de Vaucouleurs (1918-1995) was first to show in 1948 that this function best fits the galaxy light profiles, with the only difference that he held n fixed to a value of 4. 20 years later in 1968, J. L. Sérsic showed that n can have a variety of values and does not necessarily need to be 4. This profile depends on the effective radius (r_e) which is defined as the radius which contains half of the profile brightness (see Section 8.1.4 [Profile magnitude], page 135). I_e is the flux at the effective radius. The Sérsic index n is used to define the concentration of the profile within r_e and b_n is a constant dependent on n. MacArthur et al.[3] show that for $n > 0.35$, b_n can be accurately approximated using this equation:

$$b_n = 2n - \frac{1}{3} + \frac{4}{405n} + \frac{46}{25515n^2} + \frac{131}{1148175n^3} - \frac{2194697}{30690717750n^4}$$

---

[3] MacArthur, L. A., S. Courteau, and J. A. Holtzman (2003). "Structure of Disk-dominated Galaxies. I. Bulge/Disk Parameters, Simulations, and Secular Evolution". In: ApJ 582, pp. 689—722.

8.1.1.5 Sampling from a function

A pixel is the ultimate level of accuracy to gather data, we can't get any more accurate in one image, this is known as sampling in signal processing. However, the mathematical profiles which describe our models have infinite accuracy. Over a large fraction of the area of astrophysically interesting profiles (for example galaxies or PSFs), the variation of the profile over the area of one pixel is not too significant. In such cases, the elliptical radius (r_{el} of the center of the pixel can be assigned as the final value of the pixel, see Section 8.1.1.1 [Defining an ellipse], page 129).

As you approach their center, some galaxies become very sharp (their value significantly changes over one pixel's area). This sharpness increases with smaller effective radius and larger Sérsic values. Thus rendering the central value extremely inaccurate. The first method that comes to mind for solving this problem is integration. The functional form of the profile can be integrated over the pixel area in a 2D integration process. However, unfortunately numerical integration techniques also have their limitations and when such sharp profiles are needed they can become extremely inaccurate.

The most accurate method of sampling a continuous profile on a discrete space is by choosing a large number of random points within the boundaries of the pixel and taking their average value (or Monte Carlo integration). This is also, generally speaking, what happens in practice with the photons on the pixel. The number of random points can be set with `--numrandom`.

Unfortunately, repeating this Monte Carlo process would be extremely time and CPU consuming if it is to be applied to every pixel. In order to not loose too much accuracy, in MakeProfiles, the profile is built using both methods explained above. The building of the profile begins from its central pixel and continues outwards. Monte Carlo integration is first applied (which yields F_r), then the central pixel value (F_c) on the same pixel. If the fractional difference ($|F_r - F_c|/F_r$) is lower than a given tolerance level we will stop using Monte Carlo integration and only use the central pixel value.

The ordering of the pixels in this inside-out construction is based on $r = \sqrt{(i_c - i)^2 + (j_c - j)^2}$, not r_{el}, see Section 8.1.1.1 [Defining an ellipse], page 129. When the axis ratios are large (near one) this is fine. But when they are small and the object is highly elliptical, it might seem more reasonable to follow r_{el} not r. The problem is that the gradient is stronger in pixels with smaller r (and larger r_{el}) than those with smaller r_{el}. In other words, the gradient is strongest along the minor axis. So if the next pixel is chosen based on r_{el}, the tolerance level will be reached sooner and lots of pixels with large fractional differences will be missed.

Monte Carlo integration uses a random number of points. Thus, everytime you run it, by default, you will get a different distribution of points to sample within the pixel. In the case of large profiles, this will result in a slight difference of the pixels which use Monte Carlo integration each time MakeProfiles is run. To have a deterministic result, you have to fix the random number generator properties which is used to build the random distribution. This can be done by setting the `GSL_RNG_TYPE` and `GSL_RNG_SEED` environment variables and calling MakeProfiles with the `--envseed` option. To learn more about the process of generating random numbers, see Section 8.2.1.4 [Generating random numbers], page 144.

The seed values are fixed for every profile: with `--envseed`, all the profiles have the same seed and without it, each will get a different seed using the system clock (which is accurate to within one microsecond). The same seed will be used to generate a random number for all the sub-pixel positions of all the profiles. So in the former, the subpixel points checked for all the pixels undergoing Monte carlo integration in all profiles will be identical. In other words, the subpixel points in the first (closest to the center) pixel of all the profiles will be identical with each other. All the second pixels studied for all the profiles will also receive an identical (different from the first pixel) set of sub-pixel points and so on. As long as the number of random points used is large enough or the profiles are not identical, this should not cause any systematic bias.

8.1.1.6 Oversampling

The steps explained in Section 8.1.1.5 [Sampling from a function], page 133 do give an accurate representation of a profile prior to convolution. However, in an actual observation, the image is first convolved with or blurred by the atmospheric and instrument PSF in a continuous space and then it is sampled on the discrete pixels of the camera.

In order to more accurately simulate this process, the unconvolved image and the PSF are created on a finer pixel grid. In other words, the output image is a certain odd-integer multiple of the desired size, we can call this 'oversampling'. The user can specify this multiple as a command line option. The reason this has to be an odd number is that the PSF has to be centered on the center of its image. An image with an even number of pixels on each side does not have a central pixel.

The image can then be convolved with the PSF (which should also be oversampled on the same scale). Finally, image can be sub-sampled to get to the initial desired pixel size of the output image. After this, mock noise can be added as explained in the next section. This is because unlike the PSF, the noise occurs in each output pixel, not on a continuous space like all the prior steps.

8.1.2 If convolving afterwards

In case you want to convolve the image later with a given point spread function, make sure to use a larger image size. After convolution, the profiles become larger and a profile that is normally completely outside of the image might fall within it.

On one axis, if you want your final (convolved) image to be m pixels and your PSF is $2n + 1$ pixels wide, then when calling MakeProfiles, set the axis size to $m + 2n$, not m. You also have to shift all the pixel positions of the profile centers on the that axis by n pixels to the positive.

After convolution, you can crop the outer n pixels with the section crop box specification of ImageCrop: `--section=n:*-n,n:*-n` assuming your PSF is a square, see Section 6.1.2 [Crop section syntax], page 60. This will also remove all discrete Fourier transform artifacts (blurred sides) from the final image. To facilitate this shift, MakeProfiles has the options `--xshift`, `--yshift` and `--prepforconv`, see Section 8.1.5 [Invoking MakeProfiles], page 136.

8.1.3 Flux Brightness and magnitude

Astronomical data pixels are usually in units of counts[4] or electrons or either one divided by seconds. To convert from the counts to electrons, you will need to know the instrument gain. In any case, they can be directly converted to energy or energy/time using the

---

[4] Counts are also known as analog to ditigal units (ADU).

basic hardware (telescope, camera and filter) information. We will continue the discussion assuming the pixels are in units of energy/time.

The *brightness* of an object is defined as its total detected energy per time. This is simply the sum of the pixels that are associated with that detection by our detection tool for example Section 7.2 [NoiseChisel], page 112[5]. The *flux* of an object is in units of energy/time/area and for a detected object, it is defined as its brightness divided by the area used to collect the light from the source or the telescope aperture (for example in cm^2)[6]. Knowing the flux (f) and distance to the object (r), we can calculate its *luminosity*: $L = 4\pi r^2 f$. Therefore, flux and luminosity are intrinsic properties of the object, while brightness depends on our detecting tools (hardware and software). Here we will not be discussing luminosity, but brightness. However, since luminosity is the astrophysically interesting quantity, we also defined it here to avoid possible confusion between these two terms because they both have the same units.

Images of astronomical objects span over a very large range of brightness. With the Sun (as the brightest object) being roughly $2.5^{60} = 10^{24}$ times brighter than the faintest galaxies we can currently detect. Therefore discussing brightness will be very hard, and astronomers have chosen to use a logarithmic scale to talk about the brightness of astronomical objects. But the logarithm can only be usable with a unit-less and always positive value. Fortunately brightness is always positive and to remove the units we divide the brightness of the object (B) by a reference brightness (B_r). We then define the resulting logarithmic scale as *magnitude* through the following relation[7]

$$m - m_r = -2.5 \log_{10} \left(\frac{B}{B_r} \right)$$

m is defined as the magnitude of the object and m_r is the pre-defined magnitude of the reference brightness. One particularly easy condition is when $B_r = 1$. This will allow us to summarize all the hardware specific parameters discussed above into one number as the reference magnitude which is commonly known as the Zero-point[8] magnitude.

8.1.4 Profile magnitude

To find the profile brightness or its magnitude, (see Section 8.1.3 [Flux Brightness and magnitude], page 134), it is customary to use the 2D integration of the flux to infinity. However, in MakeProfiles we do not follow this idealistic approach and apply a more realistic method to find the total brightness or magnitude: the sum of all the pixels belonging to a profile within its predefined truncation radius. Note that if the truncation radius is not large enough, this can be significantly different from the total integrated light to infinity.

---

[5] If further processing is done, for example the Kron or Petrosian radii are calculated, then the detected area is not sufficient and the total area that was within the respective radius must be used.

[6] For a full object that spans over several pixels, the telescope area should be used to find the flux. However, sometimes, only the brightness per pixel is desired. In such cases this manual also *loosely* uses the term flux. This is only approximately accurate however, since while all the pixels have a fixed area, the pixel size can vary with camera on the telescope.

[7] The -2.5 factor in the definition of magnitudes is a legacy of the our ancient colleagues and in particular Hipparchus of Nicaea (190-120 BC).

[8] When $B = Br = 1$, the right side of the magnitude definition will be zero. Hence the name, "zero-point".

An integration to infinity is not a realistic condition because no galaxy extends indefinitely (important for high Sérsic index profiles), pixelation can also cause a significant difference between the actual total pixel sum value of the profile and that of integration to infinity, especially in small and high Sérsic index profiles. To be safe, you can specify a large enough truncation radius for such compact high Sérsic index profiles.

If oversampling is used then the brightness is calculated using the over-sampled image, see Section 8.1.1.6 [Oversampling], page 134 which is much more accurate. The profile is first built in an array completely bounding it with a normalization constant of unity (see Section 8.1.1.4 [Galaxies], page 132). Taking B to be the desired brightness and S to be the sum of the pixels in the created profile, every pixel is then multiplied by B/S so the sum is exactly B.

If the --individual option is called, this same array is written to a FITS file. If not, only the overlapping pixels of this array and the output image are kept and added to the output array.

8.1.5 Invoking MakeProfiles

MakeProfiles will make any number of profiles specified in a catalog either individually or in one image. The executable name is astmkprof with the following general template

```
$ astmkprof [OPTION ...] [BackgroundImage] Catalog
```

One line examples:

```
$ astmkprof background.fits catalog.txt
$ astmkprof --xcol=0 --ycol=1 catalog.txt
$ astmkprof --individual --oversample 3 -x500 -y500 catalog.txt
```

If mock galaxies are to be made, the catalog (which stores the parameters for each mock profile) is the mandatory argument. The input catalog has to be a text file formatted in a table with columns separated by space, tab or comma (,) characters. See Chapter 4 [Common behavior], page 31 for a complete explanation of some common behaviour and options in all Gnuastro programs including MakeProfiles.

If a data image file (see Section 4.1.2 [Arguments], page 32) is given, the pixels of that image are used as the background value for every pixel. The flux value of each profile pixel will be added to the pixel in that background value. In this case the values to all options relating to the output size and WCS will be ignored if specified (for example --naxis1, --naxis2 and --prepforconv) on the command line or in the configuration files. Note that --oversample will remain active even if a background image is specified.

8.1.5.1 MakeProfiles catalog

The catalog is a text file table. Its columns can be ordered in any desired manner, you can specify which columns belong to which parameters using the set of options ending with col, for example --xcol or --rcol, see Section 8.1.5.2 [MakeProfiles options], page 137.

The value for the profile center in the catalog (in the --xcol and --ycol columns) can be a floating point number so the profile center can be on any sub-pixel position. Note that pixel positions in the FITS standard start from 1 and an integer is the pixel center. So a 2D image actually starts from the position (0.5, 0.5). In MakeProfiles profile centers do not have to be in the image. Even if only one pixel of the profile within the truncation radius is within the output image, that pixel is included in the image. Profiles that are completely out of the image will not be created. You can use the output log file to see which profiles were within the image.

If PSF profiles (Moffat or Gaussian) are in the catalog and the profiles are to be built in one image (when `--individual` is not used), it is assumed they are the PSF(s) you want to convolve your created image with. So by default, they will not be built in the output image but as separate files. The sum of pixels of these separate files will also be set to unity (1) so you are ready to convolve, see Section 6.2.1.1 [Convolution process], page 66. As a summary, their position and magnitude will be ignored. This behaviour can be disabled with the `--psfinimg` option. If you want to create all the profiles separately (with `--individual`) and you want the sum of the PSF profile pixels to be unity, you have to set their magnitudes in the catalog to the zero-point magnitude and be sure that the central positions of the profiles don't have any fractional part (the PSF center has to be in the center of the pixel).

8.1.5.2 MakeProfiles options

The common options that are shared by Gnuastro programs, are fully explained in Section 4.1.4 [Common options], page 34 and are not repeated here. Since there are no image inputs, the `--hdu` option is ignored. The options can be classified into the following categories: Output, Profiles, Catalog and WCS. Below each one is reviewed.

Output:

`-x`
`--naxis1` (=INT) The number of pixels in the output image along the first FITS axis (horizontal when viewed in SAO ds9). This is before over-sampling. For example if you call MakeProfiles with `--naxis1=100 --oversample=5` (assuming no shift due for later convolution), then the final image size along the first axis will be 500. If a background image is specified, any possible value to this option is ignored.

`-y`
`--naxis2` (=INT) The number of pixels in the output image along the second FITS axis (vertical when viewed in SAO ds9), see the explanation for `--naxis1`.

`-s`
`--oversample`
(=INT) The scale to over-sample the profiles and final image. If not an odd number, will be added by one, see Section 8.1.1.6 [Oversampling], page 134. Note that this `--oversample` will remain active even if an input image is specified. If your input catalog is based on the background image, be sure to set `--oversample=1`.

`--psfinimg`
Build the possibly existing PSF profiles (Moffat or Gaussian) in the catalog into the final image. By default they are built separately so you can convolve your images with them, thus their magnitude and positions are ignored. With this option, they will be built in the final image like every other galaxy profile. To have a final PSF in your image, make a point profile where you want the PSF and after convolution it will be the PSF.

`-i`

`--individual`

> If this option is called, each profile is created in a separate FITS file within the same directory as the output and the row number of the profile (starting from zero) in the name. The file for each row's profile will be in the same directory as the final combined image of all the profiles and will have the final image's name as a suffix. So for example if the final combined image is named `./out/fromcatalog.fits`, then the first profile that will be created with this option will be named `./out/0_fromcatalog.fits`.
>
> Since each image only has one full profile out to the truncation radius the profile is centered and so, only the sub-pixel position of the profile center is important for the outputs of this option. The output will have an odd number of pixels. If there is no oversampling, the central pixel will contain the profile center. If the value to `--oversample` is larger than unity, then the profile center is on any of the central `--oversample`'d pixels depending on the fractional value of the profile center.
>
> If the fractional value is larger than half, it is on the bottom half of the central region. This is due to the FITS definition of a real number position: The center of a pixel has fractional value 0.00 so each pixel contains these fractions: .5 – .75 – .00 (pixel center) – .25 – .5.

`-m`

`--nomerged`

> Don't make a merged image. By default after making the profiles, they are added to a final image with sides of `--naxis1` and `--naxis2` if they overlap with it.

Profiles:

`-r`

`--numrandom`

> The number of random points used in the central regions of the profile, see Section 8.1.1.5 [Sampling from a function], page 133.

`-e`

`--envseed`

> Use the value to the `GSL_RNG_SEED` environment variable to generate the random Monte Carlo sampling distribution, see Section 8.1.1.5 [Sampling from a function], page 133 and Section 8.2.1.4 [Generating random numbers], page 144.

`-t`

`--tolerance`

> (=FLT) The tolerance to switch from Monte Carlo integration to the central pixel value, see Section 8.1.1.5 [Sampling from a function], page 133.

`-p`

`--tunitinp`

> The truncation column of the catalog is in units of pixels. By default, the truncation column is considered to be in units of the radial parameters of the profile (`--rcol`). Read it as 't-unit-in-p' for 'truncation unit in pixels'.

-X

--xshift   (=INT) Shift all the profiles and enlarge the image along the first FITS axis, see n in Section 8.1.2 [If convolving afterwards], page 134. This is useful when you want to convolve the image afterwards. If you are using an external PSF, be sure to oversample it to the same scale used for creating the mock images. If a background image is specified, any possible value to this option is ignored.

-Y

--yshift   (=INT) Similar to **--xshift** for the second FITS axis.

-c

--prepforconv

Shift all the profiles and enlarge the image based on half the width of the first Moffat or Gaussian profile in the catalog, considering any possible oversampling see Section 8.1.2 [If convolving afterwards], page 134. **--prepforconv** is only checked and possibly activated if **--xshift** and **--yshift** are both zero (after reading the command line and configuration files). If a background image is specified, any possible value to this option is ignored.

--magatpeak

The magnitude column in the catalog (see Section 8.1.5.1 [MakeProfiles catalog], page 136) will be used to find the brightness only for the peak profile pixel, not the full profile. Note that this is the flux of the profile's peak pixel in the final output of MakeProfiles. So beware of the oversampling, see Section 8.1.1.6 [Oversampling], page 134.

This option can be useful if you want to check a mock profile's total magnitude at various truncation radii. Without this option, no matter what the truncation radius is, the total magnitude will be the same as that given in the catalog. But with this option, the total magnitude will become brighter as you increase the truncation radius.

In sharper profiles, sometimes the accuracy of measuring the peak profile flux is more than the overall object brightness. In such cases, with this option, the final profile will be built such that its peak has the given magnitude, not the total profile.

> **CAUTION:** If you want to use this option for comparing with observations, please note that MakeProfiles does not do convolution. Unless you have deconvolved your data, your images are convolved with the instrument and atmospheric PSF, see Section 8.1.1.2 [Point Spread function], page 130. Particularly in sharper profiles, the flux in the peak pixel is strongly decreased after convolution. Also note that in such cases, besides deconvolution, you will have to set **--oversample=1** otherwise after resampling your profile with ImageWarp (see Section 6.3 [ImageWarp], page 86), the peak flux will be different.

-M

--setconsttomin

For profiles that have a constant value (no variation from pixel to pixel), set the constant value to the minimum value in the image. This is very useful if the profiles with constant value are to be used as masks. When displaying the images in a document (and inverting the images as is automatically done in ConvertType), the masked pixels will become white.

-A

--setconsttonan

>Similar to --setconsttomin, but the constant value is a NaN value. Since all Gnuastro programs treat NaN valued pixels as masked, this is useful for immediately masking some pixels in the image that have an elliptical shape without the need to creating a new mask image.

-R

--replace

>Do not add pixel values to each other, replace them. By default, when two profiles overlap, the final pixel value is the sum of all the profiles that overlap on that pixel. When this option is given, the pixels are not added but replaced by newer profiles.

>When order does matter, make sure to use this function with --numthreads=1. When multiple threads are used, the separate profiles are built asynchronously and not in order. Since order does not matter in an addition, this causes no problems by default but has to be considered when this option is given. Using multiple threads is no problem if the profiles are to be used as a mask (with --setconsttomin) since all their pixel values are the same.

-w

--circumwidth

>(=FLT) The width of the circumference if the profile is to be an elliptical circumference or annulus. See the explanations for this type of profile in --fcol.

-z

--zeropoint

>(=FLT) The zero-point magnitude of the image.

Catalog: The value to all of these options is considered to be a column number, where counting starts from zero.

--fcol (=INT) The functional form of the profile with one of the values below. Note that this value will be converted to an integer before analysis using the internal type conversion of C. So for example 2.80 will be converted to 2.

>- 0: Sérsic.
>- 1: Moffat.
>- 2: Gaussian.
>- 3: Point source (a star).
>- 4: Flat profile: all pixels have same value.
>- 5: Circumference: same value for all pixels between the truncation radius (r_t) and $r_t - w$ where w is the value to the --circumwidth. Currently this is only intended to be used for making an elliptical annulus (with a width of 1 or 2 pixels).

`--xcol`	(=INT) The center of the profiles along the first FITS axis (horizontal when viewed in SAO ds9).
`--ycol`	(=INT) The center of the profiles along the second FITS axis (vertical when viewed in SAO ds9).
`--rcol`	(=INT) The radius parameter of the profiles. Effective radius (r_e) if Sérsic, FWHM if Moffat or Gaussian.
`--ncol`	(=INT) The Sérsic index (n) or Moffat β.
`--pcol`	(=INT) The position angle (in degrees) of the profiles relative to the first FITS axis (horizontal when viewed in SAO ds9).
`--qcol`	(=INT) The axis ratio of the profiles (minor axis divided by the major axis).
`--mcol`	(=INT) The total pixelated magnitude of the profile within the truncation radius, see Section 8.1.4 [Profile magnitude], page 135.
`--tcol`	(=INT) The truncation radius of this profile. By default it is in units of the radial parameter of the profile (the value in the `--rcol` of the catalog). If `--tunitinp` is given, this value is interpreted in units of pixels (prior to oversampling) irrespective of the profile.

WCS:

`--crpix1`	(=FLT) The pixel coordinates of the WCS reference point on the first (horizontal) FITS axis (counting from 1).
`--crpix2`	(=FLT) The pixel coordinates of the WCS reference point on the second (vertical) FITS axis (counting from 1).
`--crval1`	(=FLT) The Right Ascension (RA) of the reference point.
`--crval2`	(=FLT) The Declination of the reference point.
`--resolution`	(=FLT) The resolution of the non-oversampled image in units of arcseconds/pixel.

8.1.5.3 MakeProfiles output

Besides the final merged image of all the profiles or individual profiles that can be built based on the input options, MakeProfiles will also create a log file in the current directory (where you run MockProfiles). The values for each column are explained in the first few commented (starting with # character). The log file includes the following information:

- The total magnitude of the profile in the image. This will be different from your input magnitude if the profile was not completely in the image.
- The number of pixels (in the oversampled image) which used Monte Carlo integration and not the central pixel value.
- The fraction of flux in the Monte Carlo integrated pixels.
- If an individual image was created or not.

8.2 MakeNoise

Real data are always burried in noise, therefore to finalize a simulation of real data (for example to test our observational algorithms) it is essential to add noise to the mock profiles created with MakeProfiles, see Section 8.1 [MakeProfiles], page 129. Below, the general principles and concepts to help understand how noise is quantified is discussed. MakeNoise options and argument are then discussed in Section 8.2.2 [Invoking MakeNoise], page 145.

8.2.1 Noise basics

Deep astronomical images, like those used in extragalactic studies seriously suffer from noise in the data. Generally speaking, the sources of noise in an astronomical image are photon counting noise and Instrumental noise which are discussed in detail below. We finish with a short introduction on how random numbers are generated and how you can determine the random number generator and seed value.

8.2.1.1 Photon counting noise

Thanks to the very accurate electronics used in today's detectors, this type of noise is the main cause of concern for extra galactic studies. It can generally be associate with the counting error that is known to have a Poisson distribution. The Poisson distribution is about counting. But counting is a discrete operation with only positive values, for example we can't count 3.2 or −2 of anything. We only count 0, 1, 2, 3 and so on. Therefore the Poisson distribution is also a discrete distribution, only applying to whole positive integers.

Let's assume the mean value of counting something is known. In this case, the number of electrons that are produced by photons in the CCD. Let's call this mean λ. Let's take k to represent the result of counting in one particular time we attempt to count. The probability density function of k can be written as:

$$f(k) = \frac{\lambda^k}{k!}e^{-\lambda}, \quad k \in \{0, 1, 2, 3, \ldots\}$$

Because the Poisson distribution is only applicable to positive values, it is by nature very skewed when λ is near zero. One qualitative way to imagine it is that there simply aren't enough integers smaller than λ, than there are larger integers. Therefore to accommodate all possibilities, it has to be skewed when λ is small.

But as λ becomes larger and larger, the distribution becomes more and more symmetric. One very useful property of the Poisson distribution is that the mean value is also its variance. When λ is very large, say $\lambda > 1000$, then the normal (Gaussian) distribution, see Section 8.1.1.2 [Point Spread function], page 130, is an excellent approximation of the Poisson distribution with mean $\mu = \lambda$ and standard deviation $\sigma = \sqrt{\lambda}$.

We see that the variance or dispersion of the distribution depends on the mean value, and when it is large it can be approximated with a Gaussian that only has one free parameter ($\mu = \lambda$ and $\sigma = \sqrt{\lambda}$) instead of two that it originally has.

The astronomical objects after convolution with the PSF of the instrument, lie above a certain background flux. This background flux is defined to be the average flux of a region in the image that has absolutely no objects. The physical origin of this background value is the brightness of the atmosphere or possible stray light within the imagining instrument.

It is thus an ideal definition, because in practice, what lies deep in the noise far lower than the detection limit is never known[9]. However, in a real image, a relatively large number of very faint objects can been fully buried in the noise. These undetected objects will bias the background measurement to slightly larger values. The sky value is therefore defined to be the average of the undetected regions in the image, so in an ideal case where all the objects have been detected, the sky value and background value are the same.

As longer wavelengths are used, the background value becomes more significant and also varies over a wide image field. Such variations are not currently implemented in Make-Profiles, but will be in the future. In a mock image, we have the luxury of setting the background value.

In each pixel of the canvas of pixels, the flux is the sum of contributions from various sources after convolution. Let's name this flux of the convolved sum of possibly overlapping objects, I_{nn}. nn representing 'no noise'. For now, let's assume the background is constant and represented by B. In practice the background values are larger than $\sim 1,000$ counts. Then the flux after adding noise is a random value taken from a Gaussian distribution with the following mean (μ) and standard deviation (σ):

$$\mu = B + I_{nn}, \quad \sigma = \sqrt{B + I_{nn}}$$

Since this type of noise is inherent in the objects we study, it is usually measured on the same scale as the astronomical objects, namely the magnitude system, see Section 8.1.3 [Flux Brightness and magnitude], page 134. It is then internally converted to the flux scale for further processing.

8.2.1.2 Instrumental noise

While taking images with a camera, a dark current is fed to the pixels, the variation of the value of this dark current over the pixels, also adds to the final image noise. Another source of noise is the readout noise that is produced by the electronics in the CCD that attempt to digitize the voltage produced by teh photo-electrons in the analog to digital converter. In deep extra-galactic studies these sources of noise are not as significant as the noise of the background sky. Let C represent the combined standard deviation of all these sources of noise. If only this source of noise is present, the noised pixel value would be a random value chosen from a Gaussian distribution with

$$\mu = I_{nn}, \quad \sigma = \sqrt{C^2 + I_{nn}}$$

This type of noise is completley independent of the type of objects being studied, it is completely determined by the instrument. So the flux scale (and not magnitude scale) is most commonly used for this type of noise. In practice, this value is usually reported in ADUs not flux or electron counts. The gain value of the device can be used to convert between these two, see Section 8.1.3 [Flux Brightness and magnitude], page 134.

---

[9] See the section on sky in Akhlaghi M., Ichikawa. T. 2015. Astrophysical Journal Supplement Series.

8.2.1.3 Final noised pixel value

Depending on the values you specify for B and C from the above, the final noised value for each pixel is a random value chosen from a Gaussian distribution with

$$\mu = B + I_{nn}, \quad \sigma = \sqrt{C^2 + B + I_{nn}}$$

8.2.1.4 Generating random numbers

As discussed above, to generate noise we need to make random samples of a particular distribution. So it is important to understand some general concepts regarding the generation of random numbers. For a very complete and nice introduction we strongly advise reading Donald Knuth's "The art of computer programming", volume 2, chapter 3[10]. Quoting from the GNU Scientific Library manual, "If you don't own it, you should stop reading right now, run to the nearest bookstore, and buy it"[11]!

Using only software, we can only produce what is called a psudo-random sequence of numbers. A true random number generator a hardware (let's assume we have made sure it has no systematic biases), for examle throwing dice or flipping coins (which have remained from the ancient times). More modern hardware methods use atmospheric noise, thermal noise or other types of external electromagnetic or quantum phenomena. All psudo-random number generators (software) require a seed to be the basis of the generation. The advantage of having a seed is that if you specify the same seed for multiple runs, you will get an identical sequence of random numbers which allows you to reproduce the same final noised image.

The programs in GNU Astronomy Utilities (for example MakeNoise or MakeProfiles) use the GNU Scientific Library (GSL) to generate random numbers. GSL allows the user to set the random number generator through environment variables, see Section 3.3.1.2 [Installation directory], page 25 for an introduction to environment variables. In the chapter titled "Random Number Generation" they have fully explained the various random number generators that are available (there are a lot of them!). Through the two environment variables `GSL_RNG_TYPE` and `GSL_RNG_SEED` you can sepecify the generator and its seed respectively.

If you don't specify a value for `GSL_RNG_TYPE`, GSL will use its default random number generator type. The default type is sufficient for most general applications. If no value is given for the `GSL_RNG_SEED` environment variable and you have asked Gnuastro to read the seed from the environment (through the `--envseed` option), then GSL will use the default value of each generator to give identical outputs. If you don't explicitly tell Gnuastro programs to read the seed value from the environment variable, then they will use the system time (accurate to within a microsecond) to generate (apparently random) seeds. In this manner, every time you run the program, you will get a different random number distribution.

There are two ways you can specify values for these environment variables. You can call them on the same command line for example:

---

[10] Knuth, Donald. 1998. The art of computer programming. Addison–Wesley. ISBN 0-201-89684-2

[11] For students, running to the library might be more affordable!

```
$ GSL_RNG_TYPE="taus" GSL_RNG_SEED=345 astmknoise input.fits
```

In this manner the values will only be used for this particular execution of MakeNoise. To define them for the full period of your terminal session or script length, you can use the shell's `export` command (for a script remove the `$` signs):

```
$ export GSL_RNG_TYPE="taus"
$ export GSL_RNG_SEED=345
```

The subsequent programs in the particular terminal (or script) that you ran these commands on which use of GSL's random number generators (including the Gnuastro programs) will hence forth use these values. Finally, in case you want set fixed values for these variables every time you use the GSL random number generator, you can add these two lines to your `.bashrc` startup script[12].

NOTE: If the two environment variables `GSL_RNG_TYPE` and `GSL_RNG_SEED` are defined, GSL will report them by default, even if you don't use the `--envseed` option. For example you can see the top few lines of the output of MakeProfiles:

```
$ export GSL_RNG_TYPE="taus"
$ export GSL_RNG_SEED=345
$ astmkprof catalog.txt --envseed
GSL_RNG_TYPE=taus
GSL_RNG_SEED=345
MakeProfiles started on AAA BBB DD EE:FF:GG HHH
  - 6 profiles read from catalog.txt 0.000236 seconds
  - Random number generator (RNG) type: taus
  - RNG seed for all profiles: 345
```

The first two output lines (showing the names of the environment variables) are printed by GSL before MakeProfiles actually starts generating random numbers. The Gnuastro programs will report the values they use independently, you should check them for the final values used. For example if `--envseed` is not given, `GSL_RNG_SEED` will not be used and the last line shown above will not be printed. In the case of MakeProfiles, each profile will get its own seed value.

8.2.2 Invoking MakeNoise

MakeNoise will add noise to an existing image. The executable name is `astmknoise` with the following general template

```
$ astmknoise [OPTION ...] InputImage.fits
```

One line examples:

```
$ astmknoise --background=1000 --stdadd=20 mockimage.fits
```

If actual processing is to be done, the input image is a mandatory argument. The full list of options common to all the programs in Gnuastro can be seen in Section 4.1.4 [Common options], page 34. The output will have the same type as the input image, however the

---

[12] Don't forget that if you are going to give your scripts (that use the GSL random number generator) to others you have to make sure you also tell them to set these environment variable separately. So for scripts, it is best to keep all such variable definitions within the script, even if they are within your `.bashrc`.

internal processing is done on a double precision floating point format. If the input values were integer types, then each floating point number will be rounded to the nearest integer away from zero. This might cause integer overflow if types with small ranges are used (for example images with a `BITPIX` of 8 which can only keep 256 values). This can be disabled with the `doubletype` option. The header of the output FITS file keeps all the parameters that were influential in making it. This is done for future reproducability.

`-b`

`--background`

> (=FLT) The background pixel value for the image in units of magnitudes, see Section 8.2.1.1 [Photon counting noise], page 142 and Section 8.1.3 [Flux Brightness and magnitude], page 134.

`-z`

`--zeropoint`

> (=FLT) The zeropoint magnitude used to convert the value of `--background` (in units of magnitude) to flux, see Section 8.1.3 [Flux Brightness and magnitude], page 134.

`-s`

`--stdadd`   (=FLT) The instrumental noise which is in units of flux, see Section 8.2.1.2 [Instrumental noise], page 143.

`-e`

`--envseed`

> Use the `GSL_RNG_SEED` environment variable for the seed used in the random number generator, see Section 8.2.1.4 [Generating random numbers], page 144. With this option, the output image noise is always going to be identical (or reproducable).

`-d`

`--doubletype`

> Save the output in the double precision floating point format that was used internally. This option will be most useful if the input images were of integer types.

9 Table manipulation

The FITS standard also specifies tables as a form of data that can be stored in the extensions of a FITS file. These tables can be ASCII tables or binary tables. The utilities in this section provide the tools to directly read and write to FITS tables.

The software for this section have to be added ....

10 Developing

The basic idea of GNU Astronomy Utilities is for an interested astronomer to be able to easily understand the code of any of the programs, be able to modify the code if she feels there is an improvement and finally, to be able to add new programs to the existing utilities for their own benefit, and the larger community if they are willing to share it. In short, we hope that at least from the software point of view, the "obscurantist faith in the expert's special skill and in his personal knowledge and authority" can be broken, see Section 1.2 [Science and its tools], page 2. The following software architecture can be one of the most basic and easy to understand for any interested inquirer.

First some general design choices are tackled. It is followed by a short explanation of the version controlled source. The libraries and headers in their respective directories are then explained. Later the the basic conventions for managing the code in each program to facilitate reading the code by an outside inquirer is discussed. Finally some notes on the building process are given.

10.1 Why C programming language?

Currently the programming language that is most commonly used in scientific applications is C++, and more recently Python. One of the main reasons behind this choice is that through the Object oriented programming paradigm, they offer a much higher level of abstraction. However, GNU Astronomy Utilities are written in the C programming language. The reasons can be summarized with simplicity, portability and speed. All three are very important in a scientific software.

Simplicity can best be demonstrated in a comparison of the main books of C++ and C. The "C programming language"[1] book, written by the authors of C, is only 286 pages and covers a very good fraction of the language, it has also remained unchanged from 1988. C is the main programming language of nearly all operating systems and there is no plan of any significant update. The most recent "C++ programming language"[2] book, also written by its author, on the other hand has 1366 pages and its fourth edition came out in 2013! As discussed in Section 1.2 [Science and its tools], page 2, it is very important for other scientists to be able to readily read the code of a program at their will with minimum requirements.

In C++, inheriting objects in the object oriented programming paradigm and their internal functions make the code very easy to write for the programmer who is deeply invested in those objects and understands all their relations well. But it simultaneously makes reading the program for a first time reader (a curious scientist who wants to know only how a small step was done) extremely hard. Before understanding the methods, the scientist has to invest a lot of time in understanding those objects and their relations. But in C, if only simple structures are used, all variables can be given as the basic language types for example `int`s or `float`s and their pointers to define arrays. So when an outside reader is only interested in one part of the program, that part is all they have to understand.

---

[1] Brian Kernighan, Dennis Ritchie. *The C programming language*. Prentice Hall, Inc., Second edition, 1988. It is also commonly known as K&R and is based on the ANSI C and ISO C90 standards.

[2] Bjarne Stroustrup. *The C++ programming language*. Addison-Wesley Professional; 4 edition, 2013.

Recently it is also becoming common to write scientific software in Python, or a combination of it with C or C++. Python is a high level scripting language which doesn't need compilation. It is very useful when you want to do something on the go and don't want to be halted by the troubles of compiling, linking, memory checking, etc. When the data sets are small and the job is temporary, this ability of Python is great and is highly encouraged. A very good example might be plotting, in which Python is undoubtedly one of the best.

But as the data sets increase in size and the processing becomes very complicated, the speed of Python scripts significantly decrease. So when the program doesn't change too often and is widely used in a large community mostly on large data sets (like astronomical images), using Python will waste a lot of valuable research-hours. Some use Python as a wrapper for C or C++ functions to fix the speed issue. However because such actions allow separate programs to share memory (through Python), the code in such programs tends to become extremely complicated very soon, which is contrary to the principles in Section 1.2 [Science and its tools], page 2.

Like C++, Python is object oriented, so as explained above, it needs a high level of experience with that particular program to fully understand its inner workings. To make things worse, since it is mainly for fast and on the go programming, it constantly undergoes significant changes, such that Python 2.x and Python 3.x are not compatible. Lots of research teams that invested heavily in Python 2.x cannot benefit from Python 3.x or future versions any more. Some converters are available, but since they are automatic, lots of complications might arise in the conversion. Thus, re-writing all the changes would be the only truly reliable option. If a research project begins using Python 3.x today, there is no telling how compatible their investments will be when Python 4.x or 5.x will come out. This stems from the core principles of Python, which are very useful when you look in the 'on the go' basis as described before and not future usage.

The portability of C is best demonstrated by the fact that both C++ and Python are part of the C-family of programming languages which also include Java, Perl, and many other languages. C libraries can be immediately included in C++ and with tools like SWIG[3] it is easily possible to use the C libraries in programs that are written in those languages. This will allow other scientists to use the libraries in Gnuastro with any of those languages. Gnuastro's libraries are currently static and not installed, but we are working on making them shared and installable[4]. Following that we will be working on allowing the creation of libraries in different languages at configure time[5].

The final reason was speed. This is another very important aspect of C which is not independant of simplicity (first reason discussed above). The abstractions provided by the higher-level languages (which also makes learning them harder for a newcomer) comes at the cost of speed. Since C is a low-level language[6](closer to the hardware), it is much less complex for both the human reader and the computer. The former was dicussed above in simplicity and the latter helps in making the program run more efficiently (faster). This thus allows for a closer relation between the scientist/programmer (program) and the actual

---

[3] http://swig.org/

[4] http://savannah.gnu.org/task/?13765

[5] http://savannah.gnu.org/task/?13786

[6] Low-level languages are those that directly operate the hardware like assembly languages. So C is actually a high-level language, but it can be considered the lowest-level high-level language.

data/processing. The GNU coding standards[7] also encourage the use of C over all other languages when generality of usage and "high speed" is desired.

10.2 Design philosophy

The core processing functions of each program are written mostly with the basic ISO C90 standard. We do make lots of use of the GNU additions to the C language in the GNU C Library, but these additional functions are mainly used in the user interface functions (reading your inputs and preparing them prior to or after the analysis). The actual algorithms, which most scientists would be more interested in, are much more closer to ISO C90. For this reason, the source files containing user interface code and those containing actual processing code are clearly separated, see Section 10.7 [Program source], page 154. If anything particular to the GNU C Library is used in the processing functions, it is explained in the comments in between the code.

Similar to GNU Coreutils, all the Gnuastro utilities provide very low level operations. This enables you to use the GNU Bash scripting language (which is the default in most GNU/Linux operating systems) or any other shell you might be using to operate on a large number of files or do very complex things through the creative combinations of these tools that the authors had never dreamed of. We have put a few simple examples in Chapter 2 [Tutorials], page 11.

For example all the analysis output is provided as ASCII tables which you can feed into your favorite plotting program to inspect visually. Python's Matplotlib is very useful for fast plotting of the tables to immediately check your results. If you want to include the plots in a document, you can use the PGFplots package within LATEX, no attempt is made to include such operations in Gnuastro. In short, Bash can act as a glue to connect the inputs and outputs of all these various Gnuastro utilities (and other programs) in any fashion you please.

The advantage of this architecture is that the programs become small and transparent: the starting and finishing point of every program is clearly demarcated. For nearly all operations on a modern computer, the read/write speed is very insignificant compared to the actual processing a program does. Therefore the complexity which arises from sharing memory in a large application is simply not worth the speed gain. This basic design is influenced by Eric Raymond's "The Art of Unix Programming"[8] which beautifully describes the design philosophy and practice which lead to the success of Unix-based operating systems[9].

Finally, and arguably the most important, principle of Gnuastro is this: Gnuastro is not planned to be a repository of creative programs with no clear purpose. The purpose of each program and all the major operations it does have to be very clearly documented and aligned with the general purpose of Gnuastro. Through the main management hub, we have a set of planned tasks and bugs, see Section 10.3 [Gnuastro project webpage], page 151. If you have a plan to add something and want it to be an official part of Gnuastro, please check there and if it (or something similar to it) doesn't already exist, then add it. This will notify all the developers of your intent, so potentially parallel operations do not occur

---

[7] http://www.gnu.org/prep/standards/

[8] Eric S. Raymond, 2004, *The Art of Unix Programming*, Addison-Wesley Professional Computing Series.

[9] KISS principle: Keep It Simple, Stupid!

and similar ideas can be discussed. If something similar to your idea already exists, you can contact the person in charge and join that work.

10.3 Gnuastro project webpage

Gnuastro's central management hub[10] is located on GNU Savannah[11]. It is the central software development management system for all GNU projects. Through this central hub, you can view the list of activities that the developers are engaged in, their activity on the version controlled source, and other things. Each defined activity in the development cycle is known as an 'issue' (or 'item'). An issue can be a bug (see Section 1.7 [Report a bug], page 7), or a suggested feature (see Section 1.8 [Suggest new feature], page 8) or an enhancement or generally any *one* job that is to be done. In Savannah, issues are classified into three categories or 'tracker's:

Support   This tracker is a way that (possibly anonymous) users can get in touch with the Gnuastro developers. It is a complement to the bug-gnuastro mailing list (see Section 1.7 [Report a bug], page 7). Anyone can post an issue to this tracker. The developers will not submit an issue to this list. They will only reassign the issues in this list to the other two trackers if they are valid[12]. Ideally (when the developers have time to put on Gnuastro, please don't forget that Gnuastro is a volunteer effort), there should be no open items in this tracker.

Bugs   This tracker contains all the known bugs in Gnuastro (problems with the existing tools).

Tasks   The items in this tracker contain the future plans (or new features/capabilities) that are to be added to Gnuastro.

All the trackers can be browsed by a (possibly anonymous) visitor, but to edit and comment on the Bugs and Tasks trackers, you have to be a registered Gnuastro developer. When posting an issue to a tracker, it is very important to choose the 'Category' and 'Item Group' options accurately. The first contains a list of all Gnuastro's utilities along with 'Installation', 'New utility' and 'Webpage'. The "Item Group" contains the nature of the issue, for example if it is a 'Crash' in the software (a bug), or a problem in the documentation (also a bug) or a feature request or an enhancement.

The set of horizontal links on the top of the page (Starting with 'Main' and 'Homepage' and finishing with 'News') are the easiest way to access these trackers (and other major aspects of the project) from any part of the project webpage. Hovering your mouse over them will open a drop down menu that will link you to the different things you can do on each tracker (for example, 'Submit new' or 'Browse'). When you browse each tracker, you can use the "Display Criteria" link above the list to limit the displayed issues to what you are interested in. The 'Category' and 'Group Item' (explained above) are a good starting point.

---

[10] https://savannah.gnu.org/projects/gnuastro/

[11] https://savannah.gnu.org/

[12] Some of the issues registered here might be due to a mistake on the user's side, not an actual bug in the program.

10.4 Version controlled source

The source code that is publicly distributed does not contain the revision history, it is only
the final snapshot of a stable release, ready to be configured and built. To be able to develop
successfully, the revision history of the code can be very useful, also some updates that are
not yet released might be in it. We use Git for the version control of Gnuastro. For those
who are not familiar with it, we suggest Pro Git[13]. The whole book is publicly available for
online reading and downloading. The latest version of Gnuastro can be cloned by running

```
$ git clone git://git.sv.gnu.org/gnuastro.git
```

The version controlled source code lacks the source files that we have not written or are automatically built (and included in the distributed `gnuastro-0.1.tar.gz`) for configuration
and building. There are three types of files that we have not written:

- To ensure portability for a wider range of operating systems (those that don't include
  GNU C Library, namely glibc), we have used the GNU Portability Library (Gnulib).
  Gnulib keeps a copy of all the functions in glibc and you can include them in the source
  code to replace system wide functions on other compilers.

- To test for various system architectures and compiler configurations we make use of
  the various GNU Autoconf macros in the GNU Autoconf archives[14].

- Figures in the manual. Some of them are built by LATEX (using the TikZ package) and
  some are actual rasterized images. An automatic script will do the job of drawing the
  plots with LATEX and converting all the figures into all the formats that are necessary
  for all the various formats of the manual.

To get the Gnulib and the Autoconf archives, just run these commands in any directory
you want to store them in (let's assume you are in `~/Development` directory):

```
$ pwd
/home/yourusername/Development
$ git clone git://git.sv.gnu.org/gnulib.git
$ git clone git://git.sv.gnu.org/autoconf-archive.git
```

This will download the full version controlled source of the two in separate directories.
Both these packages are regularly updated, so every once and a while you can run `$ git
pull` within them to get any possible updates. First, include all the necessary packages
from Gnulib. To do that, run the following command from within the cloned `gnuastro`
directory[15]:

```
$ ~/Development/gnulib/gnulib-tool --import --source-base=gnulib/lib \
    --with-tests --tests-base=gnulib/tests --no-vc-files math argp \
    error progname
```

One of the directories made after running `gnulib-tool` is `m4/`. We can now add the
necessary Autoconf checks from the Autoconf archives to this directory:

```
$ cp ~/Development/autoconf-archive/m4/ax_pthread.m4 ./m4/
```

---

[13] https://progit.org/

[14] http://www.gnu.org/software/autoconf-archive/

[15] Note that the `gnulib-tool` script has to be run from within the cloned Gnulib directory, it is not
insxtalled.

The last thing you have to do is to make/convert all the figures in all the necessary formats. We use LaTeX (along with TikZ) to create some of the figures and ImageMagick's `convert` tool for the conversion into various types. So make sure you have them installed. All the necessary scripts are ready and you just have to run a `make` command in the appropriate directory.

```
$ cd doc/plotsrc
$ make
$ cd ../../
```

All the necessary GNU C Library functions, Autoconf macros and manual figures are now available. The GNU build system will do the rest of the job. To generate all the other necessary files you have to run the following command in the cloned **gnuastro** directory[16].

```
$ autoreconf --install
```

Now you can easily configure, build and start hacking into the code and you have the full revision history under your fingers.

The hand-written (version controlled) code for Gnuastro, this manual and the tests are divided in the following sub-directories of the top directory. Their names are standard and descriptive enough, but a short summary is given here:

doc   The Texinfo source files for this manual and the Gnuastro webpage file(s).

include   The header files of the internal static libraries and also some other header files that are used by more than one program.

lib   The internal static libraries (only their `.c` files) are stored here. These libraries hold functions that are used by more than one program, they will not be installed.

src   This directory contains a subdirectory for every program in this version of Gnuastro. The source code for each program is placed inside each of these sub-directories to be easily separable.

tests   This directory keeps all the tests (checks) which are executed when `make check` is run.

10.5 Internal libraries

Libraries are binary (compiled) files which are not executable them selves, but once linked with other binary files, they form the building blocks of larger programs. Several functions are commonly used by all or several of the programs in Gnuastro. Therefore they are written as separate libraries so we don't have to maintain duplicate code. Such libraries are commonly referred to as convenience libraries. They are mostly to do with interaction with the outside world (of the program), for example setting up the configuration files, reading text catalogs or wrappers for CFITSIO and WCSLIB to facilitate reading and writing of FITS files. The names of the libraries are usually descriptive enough on the kind of functions they keep.

Currently these libraries are not installed along with Gnuastro, they are only statically linked to any program needing them in the build directory and remain or are deleted from

---

[16] `autoreconf` is part of GNU Autoconf and also requires GNU Automake, GNU Libtool and GNU Texinfo to create all the necessary files, see Section 10.9 [Building], page 160

there. Note that in a static link, the contents of the library are merged with the executable, so they are no longer needed after the linking (you can safely delete them after installing the executable). In the future if need be, they can also be installed so they can be used by other programs too.

10.6 Header files

The `include/` directory contains the headers for Gnuastro's internal libraries (see Section 10.5 [Internal libraries], page 153) and several header-only files in the `include` directory. Below is a list of the latter type.

`commonargs.h`

> All the programs have a common set of options, see Section 4.1.4 [Common options], page 34. Instead of including separately them and making sure they are identical in the implementation of all programs, the GNU C library's ability to merge independent argument parsers with Argp is used. This ensures that they are identical in all programs with only one file to work on. The common options and the function to parse them are thus defined in this header file. All the argument parsers in various programs are merged with this argument parser to read the user's input.

`commonparams.h`

> The structure that keeps the values of the common arguments and whether they have been set or not is defined in this header file.

`fixedstringmacros.h`

> Some strings are fixed in all the programs, only the relevant names of the packages must be put in them. The various names for each package are defined in their `main.h` source file with macros of fixed names. For example the copyright notice, or parts of the top information in the `--help` output.

`neighbors.h`

> The macros in this header find the neighbors of a pixel index using four or eight connectivity in a region of an image or the whole image.

10.7 Program source

Besides the fact that all the programs share some functions that were explained in Section 10.5 [Internal libraries], page 153, everything else about each program is completely independent. In this section the conventions used in all the program sources are explained. To easily understand the explanations in this section, it is good to open the source files of one or several of the programs in Gnuastro and inspect them as you read along.

10.7.1 Mandatory source code files

Some programs might need lots of source files and if there is no fixed convention, navigating them can become very hard for a new inquirer into the code. The following source files exist in every program's source directory (which is located in `src/progname`). For small programs, these files are enough. Larger programs will need more files. In general for writing other source files, please choose filenames that are descriptive.

main.c   Each executable has a `main` function, which is located in `main.c`. Therefore this file in any program's source code will be the starting point. No actual processing functions are to be defined in this file, the function(s) in this file are only meant to connect the most high level steps of each program. Generally, `main` will first call the top user interface function to read user input and make all the preparations. Then it will pass control to the top processing function for that program. The functions to do both these jobs must be defined in other source files.

main.h   All the major parameters which will be used in the program must be stored in a structure which is defined in `main.h`. The name of this structure is usually `prognameparams`, for example `imgcropparams`. So `#include "main.h"` will be a staple in all the source codes of the program and most of the functions. Keeping all the major parameters of a program in this structure has the major benefit that most functions will only need one argument: a pointer to this structure. This will significantly facilitate the job of the programmer, the inquirer and the computer. All the programs in Gnuastro are designed to be low-level, small and independent parts, so this structure should not get too large.

The main root structure of a program contains at least two other structures: a structure only keeping parameters for user interface functions, which is also defined in `main.h` and the `commonparams` structure which is defined in `commonparams.h`, see Section 10.6 [Header files], page 154. The main structure can become very large and since the programmer and inquirer often don't need to be confused with these parameters mixed with the actual processing parameters, they are conveniently defined in another structure which is named `uiparams` and is also defined in `main.h`. It could be defined in `ui.h` (see below) so the main functions remain completely ignorant to it, but its parameters might be needed for reporting input conditions, so it is included as part of the main program structure.

This top root structure is conveniently called `p` (short for parameters) by all the programs. The `uiparams` structure is called `up` (for user parameters) and the `commonparams` structure is called `cp`. With this convention any reader can immediately understand where to look for the definition of one parameter.

With this basic root structure, source code of functions can potentially become full of structure de-reference operators (`->`) which can make the code very unreadable. In order to avoid this, whenever a parameter is used more than a couple of times in a function, a parameter of the same type and with the same name (so it can be searched) as the desired parameter should be defined and put the value of the root structure inside of it in definition time. For example

```
char *hdu=p->cp.hdu;
int verb=p->cp.verb;
```

args.h   The argument parser structures (which are used by GNU C Library's Argp) for each program are defined in `args.h`. They are separate global variables and function definitions that will be used by Argp. We recommend going through the appropriate section in GNU C library to understand their exact meaning, although they should be descriptive and understandable enough by looking at a few of the programs.

`ui.c, ui.h`

> The user interface functions are also a unique set of functions in all the programs, so they are particularly named `ui.c` and `ui.h` in all the programs. Everything related to reading the user input arguments and options, checking the configuration files and checking the consistency of the input parameters before the actual program is run should be done in this file. Since most functions are the same, with only the internal checks and structure parameters differing, we recommend going through several of the examples and structuring your `ui.c` in a similar fashion with the rest of the programs.
>
> The most high-level function in `ui.c` is named `setparams` which accepts `int argc, char *argv[]` and a pointer to the root structure for that program, see the explanation for `main.h`. This is the function that `main` calls. The basic idea of the functions in this file is that the processing functions should need a minimum number of such checks. With this convention an inquirer who only wants to understand only one part (mostly the processing part and not user input details) of the code can easily do so. It also makes all the errors related to input appear before the processing begins which is more convenient for the user.

`progname.c`

> The main processing functions in each program which keep the function(s) that `main()` will call are in a file named `progname.c`, for example `imgcrop.c` or `noisechisel.c`. The function within these files which `main()` calls is also named after the program, for example
>
> ```
>     void
>     imgcrop(struct imgcropparams *p)
> ```
> or
> ```
>     void
>     noisechisel(struct noisechiselparams *p)
> ```
>
> In this manner, if an inquirer is interested the processing steps, they can immediately come and check this file for the first processing step without having to go through `main.c` first. In most situations, any failure in any step of the programs will result in an informative error message and an immediate abort in the program. So there is no need for return values. Under more complicated situations where a return value might be necessary, `void` will be replaced with an `int` in the examples above.

`cite.h`   This file keeps the function to be called if the user runs any of the programs with `--cite`, see Section 4.1.4.2 [Operating modes], page 35.

10.7.2 Coding conventions

Generally we try our best to follow the GNU coding standards, besides those the following conventions are adhered to until now. If new code is also added in the same manner, it would be much more easily readable by any interested astronomer (who will become familiar with it after reading once).

- It is very important that the code be easy to read by the eye. So when the order of several lines within a function does not matter (mostly when defining variables at the start of a function). You should put the lines in the order of increasing length and group the variables with similar types such that this half-pyramid of declarations becomes most visible. If the reader is interested, a simple search will show them the variable they are interested in. However, when looking through the functions or reading the separate steps of the functions, this 'order' in the declarations will make reading the rest of the function steps much more easier and pleasant to the eye.

- When ever you see that the function cannot be fully displayed (vertically) in your monitor, this is a sign that it has become too long and should be broken up into multiple functions. 40 lines is usually a good reference. When the start and end of a function are clearly visible in one glance, the function is much more easier to understand. This is most important for low-level functions (which usually define a lot of variables). Low-level functions do most of the processing, they will also be the most interesting part of a program for an inquiring astronomer. This convention is less important for higher level functions that don't define too many variables and whose only purpose is to run the lower-level functions in a specific order and with checks.

  In general you can be very liberal in breaking up the functions into smaller parts, the GNU Compiler Collection (GCC) will automatically compile the functions as inline functions when the optimizations are turned on. So you don't have to worry about decreasing the speed. By default Gnuastro will compile with the `-O3` optmization flag.

- If possible, the text files should always be at most 80 characters wide. Monitors today are certainly much wider, but with this limit, reading the functions becomes much more easier. Also for the developers, it allows multiple files (or multiple views of one file) to be displayed beside each other on wide monitors (Emacs's buffers are excellent for this capability).

  For long comments you can use press `Alt-q` in Emacs to separate them into separate lines automatically. For long literal strings, you can use the fact that in C, two strings immediately after each other are concatenated, for example `"The first part, " "and the second part."` Note the space character in the end of the first part. Since they are now separated, you can easily break a long literal string into several lines and adhere to the maximum 80 character line length policy.

- The headers required by each source file (ending with `.c`) should be defined inside of it. All the headers a program needs should not be stacked in another header to include in all source files (for example `main.h`). Although most 'professional' programmers choose the latter type, Gnuastro is primarily written for inquisitive astronomers (who are generally amateur programmers). This is very useful for readability by a first time reader. `main.h` may only include the header file(s) that define types that the main program structure needs, see `main.h` in Section 10.7 [Program source], page 154. Those particular header files that are included in `main.h` can ofcourse be ignored (not included) in separate source files.

- The headers should be classified (by an empty line) into separate groups:
  1. `#include <config.h>`: This must be the first code line (not commented or blank) in each source file. It sets macros that the GNU Portability Library (Gnulib) will use for the possible additions/modifications to C library headers.

2. The C library (or GNU C library) header files, for example `stdio.h` or `errno.h`.

3. Installed library header files, for example `cfitsio.h` or `gsl/gsl_rng.h`.

4. Gnuastro common header files, for example `fitsarrayvv.h` or `neighbors.h`, see Section 10.6 [Header files], page 154.

5. That particular program's header files, for example `main.h` and `mkprof.h`.

As much as order does not matter when you include the header of each group, sort them by length, see above.

- There should be no trailing white space in a line. To do this automatically every time you save a file in Emacs, add the following line to your `~/.emacs` file.

  ```
  (add-hook 'before-save-hook 'delete-trailing-whitespace)
  ```

- There should be no tabs in the indentation. Add the line below to your `~/.emacs` file to do this automatically:

  ```
  (setq-default indent-tabs-mode nil)
  ```

- All similar functions are separated by 5 blank lines to be easily seen to be related in a group when parsing the source code by eye. In Emacs you can use `CTRL-u 5 CTRL-o`.

- One group of functions is separated from another with 20 blank lines. In Emacs you can use `CTRL-u 20 CTRL-o`. Each group of functions has short descriptive title of the functions in that group. This title is surrounded by asterisks (∗) to make it clearly distinguishable. Such contextual grouping and clear title are very important for easily understanding the code.

10.7.3 Multithreaded programming

Most of the programs in Gnuastro utilize multi-threaded programming for the CPU intensive processing steps. This can potentially lead to a significant decrease in the running time of a program, see Section 4.3.1 [A note on threads], page 40. In terms of reading the code, you don't need to know anything about multi-threaded programming. You can simply follow the case where only one thread is to be used. In these cases, threads are not used and can be completely ignored.

At the time K&R's book was written, using threads was not common. We use POSIX threads for multi-threaded programming, defined in the `pthread.h` system wide header. There are various resources for learning to use POSIX threads, the excellent tutorial from Lawrence Livermore National Laboratory[17] is a very good start. The book 'Advanced programming in the Unix environment'[18], by Richard Stevens and Stephen Rago, Addison-Wesley, 2013 (Third edition) also has two chapters explaining the POSIX thread constructs which can be very helpful.

An alternative to POSIX threads was OpenMP, but POSIX threads are low level, allowing much more control, while being easier to understand, see Section 10.1 [Why C programming language?], page 148. All the situations where threads are used are completely independent with minimal need of coordination between the threads. Such problems are known as "embarrassingly parallel" problems. They are some of the simplest problems

---

[17] https://computing.llnl.gov/tutorials/pthreads/

[18] Don't let the title scare you! The two chapters on Multi-threaded programming are very self sufficient and don't need any more knowledge than K&R.

to solve with threads and also the ones that benefit most from threads, see the LLNL introduction[19].

10.7.4 Documentation

Documentation (this manual) is an integral part of Gnuastro. Documentation is not considered a separate project. So, no change is considered valid for implementation unless the respective parts of the manual have also been updated. The following procedure can be a good suggestion to take when you have a new idea and are about to start implementing it.

The steps below are not a requirement, the important thing is that when you send the program to be included in Gnuastro, the manual and the code have to both be fully up-to-date and compatible and the purpose should be very clearly explained. You can follow any path you choose to do this, the following path was what we found to be most successful during the initial design and implementation steps of Gnuastro.

1. Edit the manual and fully explain your desired change, such that your idea is completely embedded in the general context of the manual with no sence of discontinuity for a first time reader. This will allow you to plan the idea much more accurately and in the general context of Gnuastro or a particular program. Later on, when you are coding, this general context will significantly help you as a road-map.

   A very important part of this process is the program introduction, which explains the purposes of the program. Before actually starting to code, explain your idea's purpose thoroughly in the start of the program section you wish to add or edit. While actually writing its purpose for a new reader, you will probably get some very valuable ideas that you hadn't thought of before, this has occured several times during the creation of Gnuastro. If an introduction already exists, embed or blend your idea's purpose with the existing purposes. We emphasize that doing this is equally useful for you (as the programmer) as it is useful for the user (reader). Recall that the purpose of a program is very important, see Section 10.2 [Design philosophy], page 150.

   As you have already noticed for every program, it is very important that the basics of the science and technique be explained in separate subsections prior to the 'Invoking Programname' subsection. If you are writing a new program or your addition involves a new concept, also include such subsections and explain the concepts so a person completely unfamiliar with the concepts can get a general initial understanding. You don't have to go deep into the details, just enough to get an interested person (with absolutely no background) started. If you feel you can't do that, then you have probably not understood the concept your self! Have in mind that your only limitation in length is the fatigue of the reader after reading a long text, nothing else.

   It might also help if you start implementing your idea in the 'Invoking ProgramName' subsection (explaining the options and arguments you have in mind) at this stage too. Actually starting to write it here will really help you later when you are coding.

2. After you have finished adding your initial intended plan to the manual, then start coding your change or new program within the Gnuastro source files. While you are coding, you will notice that somethings should be different from what you wrote in the manual (your initial plan). So correct them as you are actually coding.

---

[19] https://computing.llnl.gov/tutorials/parallel_comp/

3. In the end, read the section in the manual that you edited completely and see if you didn't miss any change in the coding and to see if the context is fairly continuous for a first time reader (who hasn't seen the manual or had known of Gnuastro before you made your change).

10.8 Test scripts

As explained in Section 3.3.2 [Tests], page 28, for every program some simple tests are written to check the various independent features of the program. All the tests are placed in the `gnuastro-0.1/tests` directory, let's call it `TESTdir`. There is one script (`prepconf.sh`) in this folder and several `Makefiles`. The script is the first 'test' that will be run. It will copy all the configuration files from the various directories to a `.gnuastro` directory which it will make so the various tests can set the default values.

For each program, the tests are placed inside directories with the program name. Each test is written as a shell script. The last line of this script is the test which runs the program with certain parameters. The return value of this script determines the fate of the test, see the "Support for test suites" chapter of the Automake manual for a very nice and complete explanation. In every script, two variables are defined at first: `prog` and `execname`. The first specifies the program name and the second the location of the executable.

The most important thing to have in mind about all the test scripts is that they are run from inside the `TESTdir` directory in the "build tree". Which can be different from the directory they are stored in (known as the "source tree"). This distinction is made by GNU Autoconf and Automake (which configure, build and install Gnuastro) so that you can install the program even if you don't have write access to the directory keeping the source files. See the "Parallel build trees (a.k.a VPATH builds)" in the Automake manual for a nice explanation.

Because of this, any possible data that was not generated by other tests (and is thus in the build tree), for example the catalogs in ImageCrop tests, has a `$topsrc` prefix instead of `../` for the build three. This `$topsrc` variable points to the source tree where the script can find the source data (it is defined in `TESTdir/Makefile.am`). The executables and other test products were built in the build tree (where they are being run), so they don't need to be prefixed with that variable. This is also true for images or files that were produced by other tests.

10.9 Building

To build the various programs and libraries in Gnuastro, the GNU build system is used which defines the steps in Section 1.1 [Quick start], page 1. It consists of GNU Autoconf and GNU Automake and GNU Libtool which are collectively known as GNU Autotools. They provide a very portable system to check the environment a program is to be installed on prior to compiling and set the compilation conditions based on the particular user. They also make installing everything in their standard places very easy for the programmer. Most of the small caps files that you see in the top `gnuastro-0.1/` directory are created by these three tools.

By default all the programs are compiled with optimization flags for increased speed. A side effect is that valuable debugging information is lost. To compile with the debugging flag set on (and no optimization) you can add the following options to configure:

```
$ ./configure CFLAGS="-g -OO"
```

In order to understand the building process, you can go through the Autoconf, Automake and Libtool manuals, like all GNU manuals they provide both a great tutorial and technical documentation. The "A small Hello World" section in Automake's manual (in chapter 2) can be a good starting guide after you have read the introductions of both. To get a good understand of how these three operate separately yet the codes are all mixed, there is a great tutorial book[20] which you can get you started off.

10.10 After making changes

After you have made your your changes/additions, please take the following steps:

1. Write test(s) in the **tests/progname/** directory to test the change(s)/addition(s) you have made. Then add their file names to **tests/Makefile.am**. And run **$ make check** to make sure everything is working correctly.

2. Make sure the manual is completely up to date with your changes, see Section 10.7.4 [Documentation], page 159.

3. If you have *changed* anything in the program, add it to the **ChangeLog** (a file in the top source code directory). **ChangeLog** has a specific format defined by the GNU coding standards. The easiest way to add an entry to it is through Emacs: by pressing **CTRL-x 4 a** within the places that you have changed. Note that if you have only added something, there is no need to include it in **ChangeLog**.

4. Finally, to make sure everything will run and is checked correctly, run

   ```
   $ make distcheck
   ```

   This command will create a distribution file (ending with **.tar.gz**) and try to compile it in the most general cases, then it will run the tests on what it has built in its own mini-environment. If **$ make distcheck** finishes successfully, then you are safe to send your changes to us to implement or for your own purposes.

10.11 Contributing to Gnuastro

After you have checked your changes (see Section 10.10 [After making changes], page 161), you might want to share them with the larger community. In this section we describe the Commit guidelines and the general workflow that is currently planned for Gnuastro. This workflow is currently mostly borrowed from the general recommendations of Git[21] and GitHub. But since Gnuastro is under heavy development currently, these might change and evolve to better suite our needs.

10.11.1 Commit guidelines

We strive to follow these standards for the commit messages in Gnuastro.

Commit title

The commits have to start with one short descriptive title. This will help when getting a log output of all the commits, since most emulated command line terminals are about 80 characters wide, it is best for this title to be less

---

[20] https://www.sourceware.org/autobook/

[21] https://github.com/git/git/blob/master/Documentation/SubmittingPatches

than about 60 (or even 50) characters to also allow for the commit labels and branches. The title should also not finish with any full-stops or periods.

Commit body

The body of the commit message should be separated with the title by one empty line. The body should be very descriptive. Try to explain the committed contents as best as you can. Try to remember that the people (or even yourself) who will be reading this later don't have the background that you currently have. So be very descriptive and explain as much as possible: what the bug/task was, justify the way you fixed it and discuss other possible solutions that you might not have included. For the last item, it is best to discuss them thoroughly as comments in the appropriate section of the code, but only give a short summary in the commit message.

A reference to a bug or task ID on Savannah is necessary when the commit has to do with such issues. If the commit is a fix to a bug, start the commit body with "`Fixed bug #ID:`" and if it is a task, "`Finished task #ID:`" followed by a description of what the bug/task was. Although giving URL references in the description can be very good to help others understand the issue, it is best to also accompany each such URL with a description of the relevant point within it (assume that the reader doesn't have internet access when reading the commit message).

10.11.2 Production workflow

Fortunately Pro Git has done a wonderful job in explaining the different workflows in Chapter 5[22] and in particular the "Integration-Manager Workflow" explained there. The implementation of this workflow is nicely explained in Section 5.2[23] under "Forked-Public-Project". Upon a release, a tag will be put on the appropriate commit in master marking the release version. So, whomever clones Gnuastro will automatically go onto the ready to be used master branch that contains bug-fixes or additions that might not yet be released. Anything on the master branch should always be tested and ready to be used.

As fully elaborated in Pro Git, there are two methods for you to contribute to Gnuastro in this workflow:

1. You can send commit patches by email as fully explained in Pro Git. This is good for your first few contributions, if you would like to get more heavily involved in Gnuastro's development, then you can try the next solution.

2. You can have your own forked copy of Gnuastro on any hosting site you like (GitHub, GitLab, BitBucket, or etc) and inform us when your changes are ready so we merge them in Gnuastro. This is more suited for people who commonly contribute to the code.

   Just please follow the feature branch methodology: create a new branch for every feature you want to add from the most recent master commit in the main repo, when your work is done, inform us of your repository and the branch. After testing it and if everything is working fine, we will merge it into the main repository. If you have

---

[22] http://git-scm.com/book/en/v2/Distributed-Git-Distributed-Workflows

[23] http://git-scm.com/book/en/v2/Distributed-Git-Contributing-to-a-Project

made temporary branches within that branch, it is best to rebase them into one branch before you inform us.

In both cases, your commits (with your name and information) will be preserved and your contributions will thus be fully recorded in the history of Gnuastro. Needless to say that in such cases, be sure to follow the bug or task trackers and contact us before hand so you don't do something that someone else is already working on. In that case, you can get in touch with them and help the job go on faster, see Section 10.3 [Gnuastro project webpage], page 151.

Appendix A GNU Astronomy Utilities list

GNU Astronomy Utilities 0.1, contains the following programs. They are sorted in alphabetical order and followed by their version number. A short description is provided for each program which starts with the executable names in parenthesis, see Section 1.4 [Naming convention], page 4. Throughout this manual, they are ordered based on their context, please see the manual contents for contextual ordering.

ConvertType 0.1

> (`astconvertt`) Convert astronomical data files (FITS or IMH) to and from several other standard image and data formats, for example JPEG, EPS or PDF (Section 5.2 [ConvertType], page 51).

Convolve 0.1

> (`astconvolve`) Convolve (blur or smooth) data with a given kernel (Section 6.2 [Convolve], page 65).

Header 0.1

> (`astheader`) Print and manipulate the header data of a FITS file (see Section 5.1 [Header], page 48).

ImageCrop 0.1

> (`astimgcrop`) Crop region(s) from an image and stitch several images if necessary. Inputs can be in pixel coordinates or world coordinates (Section 6.1 [ImageCrop], page 58).

ImageStatistics 0.1

> (`astimgstat`) Get pixel statistics and save histogram and cumulative frequency plots (Section 7.1 [ImageStatistics], page 104).

ImageWarp 0.1

> (`astimgwarp`) Warp image to new pixel grid (Section 6.3 [ImageWarp], page 86).

MakeCatalog 0.1

> (`astmkcatalog`) Make catalog of labeled image (Section 7.3 [MakeCatalog], page 121).

MakeNoise 0.1

> (`astmknoise`) Make (add) noise to an image (Section 8.2 [MakeNoise], page 142).

MakeProfiles 0.1

> (`astmkprof`) Make mock profiles in image (Section 8.1 [MakeProfiles], page 129).

NoiseChisel 0.1

> (`astnoisechisel`) Detect and segment signal in noise (Section 7.2 [NoiseChisel], page 112).

SubtractSky 0.1

> (`astsubtractsky`) Find and subtract sky value by comparing the mode and median on a mesh grid (Section 6.4 [SubtractSky], page 93).

Appendix B Other useful software

In this appendix the installation of programs and libraries that are not direct Gnuastro dependencies are discussed. However they can be useful for working with Gnuastro.

B.1 SAO ds9

SAO ds9[1] is not a requirement of Gnuastro, it is a FITS image viewer. So to check your inputs and outputs, it is one of the best options. Like the other packages, it might already be available in your distribution's repositories. It is already pre-compiled in the download section of its webpage. Once you download it you can unpack and install (move it to a system recognized directory) with the following commands (`x.x.x` is the version number):

```
$ tar -zxvf ds9.linux64.x.x.x.tar.gz
$ sudo mv ds9 /usr/local/bin
```

Once you run it, there might be a complaint about the Xss library, which you can find in your distribution package management system. You might also get an `XPA` related error. In this case, you have to add the following line to your `~/.bashrc` and `~/.profile` file (you will have to log out and back in again for the latter):

```
export XPA_METHOD=local
```

B.1.1 Viewing multiextension FITS images

The FITS definition allows for multiple extensions inside a FITS file, each extension can have a completely independent data set inside of it. If you ordinarily open a multi-extension FITS file with SAO ds9, for example by double clicking on the file or running `$ds9 foo.fits`, SAO ds9 will only show you the first extension. To be able to switch between the extensions you have to follow these menus in the SAO ds9 window: File→Open Other→Open Multi Ext Cube and then choose the Multi extension FITS file in your computer's file structure.

The method above is a little tedious to do every time you want view a multi-extension FITS file. Fortunately SAO ds9 also provides options that you can use to specify a particular behavior. One of those options is `-mecube` which opens a FITS image as a multi-extension data cube. So on the command line, if you run `$ds9 -mecube foo.fits` a small window will also be opened, which allows you to switch between the image extensions that `foo.fits` might have. If `foo.fits` only consists of one extension, then SAO ds9 will open as usual.

Just to avoid confusion, note that SAO ds9 does not follow the GNU style of separating long and short options as explained in Section 4.1.1 [Arguments and options], page 31. In the GNU style, this 'long' option should have been called like `--mecube`, but SAO ds9 does follow those conventions and has its own.

It is really convenient if you set ds9 to always run with the `-mecube` option on your graphical display. On GNOME 3 (the most popular graphic user interface for GNU/Linux systems) you can do this by taking the following steps:

- Open your favorite text editor and put the following text in a file that ends with `.desktop`, for example `saods9.desktop`. The file is very descriptive.

---

[1] http://ds9.si.edu/

```
[Desktop Entry]
Type=Application
Version=1.0
Name=SAO ds9
Comment=View FITS images
Exec=ds9 -mecube %f
Terminal=false
Categories=Graphic;FITS;
```

- Copy this file into your local (user) applications directory:

      $ cp saods9.desktop ~/.local/share/applications/

  In case you don't have the directory, you can make it your self:

      $ mkdir -p ~/.local/share/applications/

- The steps above will add SAO ds9 as one of your applications. To make it default for every time you click on a FITS file. Right click on a FITS file and select "Open With", then go into "Other Application..." and choose "SAO ds9".

In case you are using GNOME 2 you can take the following steps: right click on a FITS file and choose Properties→Open With→Add button. A list of applications will show up, ds9 might already be present in the list, but don't choose it because it will run with no options. Below the list is an option "Use a custom command". Click on it and write the following command: **ds9 -mecube** in the box and click "Add". Then finally choose the command you just added as the default and click the "Close" button.

B.2 PGPLOT

PGPLOT is a package for making plots in C. It is not directly needed by Gnuastro, but can be used by WCSLIB, see Section 3.1.3 [WCSLIB], page 22. As explained in Section 3.1.3 [WCSLIB], page 22, you can install WCSLIB without it too. It is very old (the most recent version was released early 2001!), but remains one of the main packages for plotting directly in C. WCSLIB uses this package to make plots if you want it to make plots. If you are interested you can also use it for your own purposes.

If you want your plotting codes in between your C program, PGPLOT is currently one of your best options. The recommended alternative to this method is to get the raw data for the plots in text files and input them into any of the various more modern and capable plotting tools separately, for example the Matplotlib library in Python or PGFplots in LaTeX. This will also significantly help code readability. Let's get back to PGPLOT for the sake of WCSLIB. Installing it is a little tricky (mainly because it is so old!).

You can download the most recent version from the FTP link in its webpage[2]. You can unpack it with the **tar -vxzf** command. Let's assume the directory you have unpacked it to is **PGPLOT**, most probably it is: **/home/username/Downloads/pgplot/**. open the **drivers.list** file:

      $ gedit drivers.list

Remove the ! for the following lines and save the file in the end:

---

[2] http://www.astro.caltech.edu/~tjp/pgplot/

```
PSDRIV 1 /PS
PSDRIV 2 /VPS
PSDRIV 3 /CPS
PSDRIV 4 /VCPS
XWDRIV 1 /XWINDOW
XWDRIV 2 /XSERVE
```

Don't choose GIF or VGIF, there is a problem in their codes.

Open the PGPLOT/sys_linux/g77_gcc.conf file:

```
$ gedit PGPLOT/sys_linux/g77_gcc.conf
```

change the line saying: FCOMPL="g77" to FCOMPL="gfortran", and save it. This is a very important step during the compilation of the code if you are in GNU/Linux. You now have to create a folder in /usr/local, don't forget to replace PGPLOT with your unpacked address:

```
$ su
# mkdir /usr/local/pgplot
# cd /usr/local/pgplot
# cp PGPLOT/drivers.list ./
```

To make the Makefile, type the following command:

```
# PGPLOT/makemake PGPLOT linux g77_gcc
```

It should finish by saying: Determining object file dependencies. You have done the hard part! The rest is easy: run these three commands in order:

```
# make
# make clean
# make cpg
```

Finally you have to place the position of this directory you just made into the LD_LIBRARY_PATH environment variable and define the environment variable PGPLOT_DIR. To do that, you have to edit your .bashrc file:

```
$ cd ~
$ gedit .bashrc
```

Copy these lines into the text editor and save it:

```
PGPLOT_DIR="/usr/local/pgplot/"; export PGPLOT_DIR
LD_LIBRARY_PATH=$LD_LIBRARY_PATH:/usr/local/pgplot/
export LD_LIBRARY_PATH
```

You need to log out and log back in again so these definitions take effect. After you logged back in, you want to see the result of all this labor, right? Tim Pearson has done that for you, create a temporary folder in your home directory and copy all the demonstration files in it:

```
$ cd ~
$ mkdir temp
$ cd temp
$ cp /usr/local/pgplot/pgdemo* ./
$ ls
```

You will see a lot of pgdemoXX files, where XX is a number. In order to execute them type the following command and drink your coffee while looking at all the beautiful plots! You are now ready to create your own.

```
$ ./pgdemoXX
```

Appendix C GNU Free Documentation License

Version 1.3, 3 November 2008

Copyright © 2000, 2001, 2002, 2007, 2008 Free Software Foundation, Inc.
http://fsf.org/

Everyone is permitted to copy and distribute verbatim copies
of this license document, but changing it is not allowed.

0. PREAMBLE

The purpose of this License is to make a manual, textbook, or other functional and useful document *free* in the sense of freedom: to assure everyone the effective freedom to copy and redistribute it, with or without modifying it, either commercially or non-commercially. Secondarily, this License preserves for the author and publisher a way to get credit for their work, while not being considered responsible for modifications made by others.

This License is a kind of "copyleft", which means that derivative works of the document must themselves be free in the same sense. It complements the GNU General Public License, which is a copyleft license designed for free software.

We have designed this License in order to use it for manuals for free software, because free software needs free documentation: a free program should come with manuals providing the same freedoms that the software does. But this License is not limited to software manuals; it can be used for any textual work, regardless of subject matter or whether it is published as a printed book. We recommend this License principally for works whose purpose is instruction or reference.

1. APPLICABILITY AND DEFINITIONS

This License applies to any manual or other work, in any medium, that contains a notice placed by the copyright holder saying it can be distributed under the terms of this License. Such a notice grants a world-wide, royalty-free license, unlimited in duration, to use that work under the conditions stated herein. The "Document", below, refers to any such manual or work. Any member of the public is a licensee, and is addressed as "you". You accept the license if you copy, modify or distribute the work in a way requiring permission under copyright law.

A "Modified Version" of the Document means any work containing the Document or a portion of it, either copied verbatim, or with modifications and/or translated into another language.

A "Secondary Section" is a named appendix or a front-matter section of the Document that deals exclusively with the relationship of the publishers or authors of the Document to the Document's overall subject (or to related matters) and contains nothing that could fall directly within that overall subject. (Thus, if the Document is in part a textbook of mathematics, a Secondary Section may not explain any mathematics.) The relationship could be a matter of historical connection with the subject or with related matters, or of legal, commercial, philosophical, ethical or political position regarding them.

The "Invariant Sections" are certain Secondary Sections whose titles are designated, as being those of Invariant Sections, in the notice that says that the Document is released under this License. If a section does not fit the above definition of Secondary then it is not allowed to be designated as Invariant. The Document may contain zero Invariant Sections. If the Document does not identify any Invariant Sections then there are none.

The "Cover Texts" are certain short passages of text that are listed, as Front-Cover Texts or Back-Cover Texts, in the notice that says that the Document is released under this License. A Front-Cover Text may be at most 5 words, and a Back-Cover Text may be at most 25 words.

A "Transparent" copy of the Document means a machine-readable copy, represented in a format whose specification is available to the general public, that is suitable for revising the document straightforwardly with generic text editors or (for images composed of pixels) generic paint programs or (for drawings) some widely available drawing editor, and that is suitable for input to text formatters or for automatic translation to a variety of formats suitable for input to text formatters. A copy made in an otherwise Transparent file format whose markup, or absence of markup, has been arranged to thwart or discourage subsequent modification by readers is not Transparent. An image format is not Transparent if used for any substantial amount of text. A copy that is not "Transparent" is called "Opaque".

Examples of suitable formats for Transparent copies include plain ASCII without markup, Texinfo input format, LaTeX input format, SGML or XML using a publicly available DTD, and standard-conforming simple HTML, PostScript or PDF designed for human modification. Examples of transparent image formats include PNG, XCF and JPG. Opaque formats include proprietary formats that can be read and edited only by proprietary word processors, SGML or XML for which the DTD and/or processing tools are not generally available, and the machine-generated HTML, PostScript or PDF produced by some word processors for output purposes only.

The "Title Page" means, for a printed book, the title page itself, plus such following pages as are needed to hold, legibly, the material this License requires to appear in the title page. For works in formats which do not have any title page as such, "Title Page" means the text near the most prominent appearance of the work's title, preceding the beginning of the body of the text.

The "publisher" means any person or entity that distributes copies of the Document to the public.

A section "Entitled XYZ" means a named subunit of the Document whose title either is precisely XYZ or contains XYZ in parentheses following text that translates XYZ in another language. (Here XYZ stands for a specific section name mentioned below, such as "Acknowledgements", "Dedications", "Endorsements", or "History".) To "Preserve the Title" of such a section when you modify the Document means that it remains a section "Entitled XYZ" according to this definition.

The Document may include Warranty Disclaimers next to the notice which states that this License applies to the Document. These Warranty Disclaimers are considered to be included by reference in this License, but only as regards disclaiming warranties: any other implication that these Warranty Disclaimers may have is void and has no effect on the meaning of this License.

2. VERBATIM COPYING

You may copy and distribute the Document in any medium, either commercially or noncommercially, provided that this License, the copyright notices, and the license notice saying this License applies to the Document are reproduced in all copies, and that you add no other conditions whatsoever to those of this License. You may not use technical measures to obstruct or control the reading or further copying of the copies you make or distribute. However, you may accept compensation in exchange for copies. If you distribute a large enough number of copies you must also follow the conditions in section 3.

You may also lend copies, under the same conditions stated above, and you may publicly display copies.

3. COPYING IN QUANTITY

If you publish printed copies (or copies in media that commonly have printed covers) of the Document, numbering more than 100, and the Document's license notice requires Cover Texts, you must enclose the copies in covers that carry, clearly and legibly, all these Cover Texts: Front-Cover Texts on the front cover, and Back-Cover Texts on the back cover. Both covers must also clearly and legibly identify you as the publisher of these copies. The front cover must present the full title with all words of the title equally prominent and visible. You may add other material on the covers in addition. Copying with changes limited to the covers, as long as they preserve the title of the Document and satisfy these conditions, can be treated as verbatim copying in other respects.

If the required texts for either cover are too voluminous to fit legibly, you should put the first ones listed (as many as fit reasonably) on the actual cover, and continue the rest onto adjacent pages.

If you publish or distribute Opaque copies of the Document numbering more than 100, you must either include a machine-readable Transparent copy along with each Opaque copy, or state in or with each Opaque copy a computer-network location from which the general network-using public has access to download using public-standard network protocols a complete Transparent copy of the Document, free of added material. If you use the latter option, you must take reasonably prudent steps, when you begin distribution of Opaque copies in quantity, to ensure that this Transparent copy will remain thus accessible at the stated location until at least one year after the last time you distribute an Opaque copy (directly or through your agents or retailers) of that edition to the public.

It is requested, but not required, that you contact the authors of the Document well before redistributing any large number of copies, to give them a chance to provide you with an updated version of the Document.

4. MODIFICATIONS

You may copy and distribute a Modified Version of the Document under the conditions of sections 2 and 3 above, provided that you release the Modified Version under precisely this License, with the Modified Version filling the role of the Document, thus licensing distribution and modification of the Modified Version to whoever possesses a copy of it. In addition, you must do these things in the Modified Version:

A. Use in the Title Page (and on the covers, if any) a title distinct from that of the Document, and from those of previous versions (which should, if there were any, be listed in the History section of the Document). You may use the same title as a previous version if the original publisher of that version gives permission.

B. List on the Title Page, as authors, one or more persons or entities responsible for authorship of the modifications in the Modified Version, together with at least five of the principal authors of the Document (all of its principal authors, if it has fewer than five), unless they release you from this requirement.

C. State on the Title page the name of the publisher of the Modified Version, as the publisher.

D. Preserve all the copyright notices of the Document.

E. Add an appropriate copyright notice for your modifications adjacent to the other copyright notices.

F. Include, immediately after the copyright notices, a license notice giving the public permission to use the Modified Version under the terms of this License, in the form shown in the Addendum below.

G. Preserve in that license notice the full lists of Invariant Sections and required Cover Texts given in the Document's license notice.

H. Include an unaltered copy of this License.

I. Preserve the section Entitled "History", Preserve its Title, and add to it an item stating at least the title, year, new authors, and publisher of the Modified Version as given on the Title Page. If there is no section Entitled "History" in the Document, create one stating the title, year, authors, and publisher of the Document as given on its Title Page, then add an item describing the Modified Version as stated in the previous sentence.

J. Preserve the network location, if any, given in the Document for public access to a Transparent copy of the Document, and likewise the network locations given in the Document for previous versions it was based on. These may be placed in the "History" section. You may omit a network location for a work that was published at least four years before the Document itself, or if the original publisher of the version it refers to gives permission.

K. For any section Entitled "Acknowledgements" or "Dedications", Preserve the Title of the section, and preserve in the section all the substance and tone of each of the contributor acknowledgements and/or dedications given therein.

L. Preserve all the Invariant Sections of the Document, unaltered in their text and in their titles. Section numbers or the equivalent are not considered part of the section titles.

M. Delete any section Entitled "Endorsements". Such a section may not be included in the Modified Version.

N. Do not retitle any existing section to be Entitled "Endorsements" or to conflict in title with any Invariant Section.

O. Preserve any Warranty Disclaimers.

If the Modified Version includes new front-matter sections or appendices that qualify as Secondary Sections and contain no material copied from the Document, you may at your option designate some or all of these sections as invariant. To do this, add their titles to the list of Invariant Sections in the Modified Version's license notice. These titles must be distinct from any other section titles.

You may add a section Entitled "Endorsements", provided it contains nothing but endorsements of your Modified Version by various parties—for example, statements of peer review or that the text has been approved by an organization as the authoritative definition of a standard.

You may add a passage of up to five words as a Front-Cover Text, and a passage of up to 25 words as a Back-Cover Text, to the end of the list of Cover Texts in the Modified Version. Only one passage of Front-Cover Text and one of Back-Cover Text may be added by (or through arrangements made by) any one entity. If the Document already includes a cover text for the same cover, previously added by you or by arrangement made by the same entity you are acting on behalf of, you may not add another; but you may replace the old one, on explicit permission from the previous publisher that added the old one.

The author(s) and publisher(s) of the Document do not by this License give permission to use their names for publicity for or to assert or imply endorsement of any Modified Version.

5. COMBINING DOCUMENTS

You may combine the Document with other documents released under this License, under the terms defined in section 4 above for modified versions, provided that you include in the combination all of the Invariant Sections of all of the original documents, unmodified, and list them all as Invariant Sections of your combined work in its license notice, and that you preserve all their Warranty Disclaimers.

The combined work need only contain one copy of this License, and multiple identical Invariant Sections may be replaced with a single copy. If there are multiple Invariant Sections with the same name but different contents, make the title of each such section unique by adding at the end of it, in parentheses, the name of the original author or publisher of that section if known, or else a unique number. Make the same adjustment to the section titles in the list of Invariant Sections in the license notice of the combined work.

In the combination, you must combine any sections Entitled "History" in the various original documents, forming one section Entitled "History"; likewise combine any sections Entitled "Acknowledgements", and any sections Entitled "Dedications". You must delete all sections Entitled "Endorsements."

6. COLLECTIONS OF DOCUMENTS

You may make a collection consisting of the Document and other documents released under this License, and replace the individual copies of this License in the various documents with a single copy that is included in the collection, provided that you follow the rules of this License for verbatim copying of each of the documents in all other respects.

You may extract a single document from such a collection, and distribute it individually under this License, provided you insert a copy of this License into the extracted document, and follow this License in all other respects regarding verbatim copying of that document.

7. AGGREGATION WITH INDEPENDENT WORKS

A compilation of the Document or its derivatives with other separate and independent documents or works, in or on a volume of a storage or distribution medium, is called an "aggregate" if the copyright resulting from the compilation is not used to limit the legal rights of the compilation's users beyond what the individual works permit. When the Document is included in an aggregate, this License does not apply to the other works in the aggregate which are not themselves derivative works of the Document.

If the Cover Text requirement of section 3 is applicable to these copies of the Document, then if the Document is less than one half of the entire aggregate, the Document's Cover Texts may be placed on covers that bracket the Document within the aggregate, or the electronic equivalent of covers if the Document is in electronic form. Otherwise they must appear on printed covers that bracket the whole aggregate.

8. TRANSLATION

Translation is considered a kind of modification, so you may distribute translations of the Document under the terms of section 4. Replacing Invariant Sections with translations requires special permission from their copyright holders, but you may include translations of some or all Invariant Sections in addition to the original versions of these Invariant Sections. You may include a translation of this License, and all the license notices in the Document, and any Warranty Disclaimers, provided that you also include the original English version of this License and the original versions of those notices and disclaimers. In case of a disagreement between the translation and the original version of this License or a notice or disclaimer, the original version will prevail.

If a section in the Document is Entitled "Acknowledgements", "Dedications", or "History", the requirement (section 4) to Preserve its Title (section 1) will typically require changing the actual title.

9. TERMINATION

You may not copy, modify, sublicense, or distribute the Document except as expressly provided under this License. Any attempt otherwise to copy, modify, sublicense, or distribute it is void, and will automatically terminate your rights under this License.

However, if you cease all violation of this License, then your license from a particular copyright holder is reinstated (a) provisionally, unless and until the copyright holder explicitly and finally terminates your license, and (b) permanently, if the copyright holder fails to notify you of the violation by some reasonable means prior to 60 days after the cessation.

Moreover, your license from a particular copyright holder is reinstated permanently if the copyright holder notifies you of the violation by some reasonable means, this is the first time you have received notice of violation of this License (for any work) from that copyright holder, and you cure the violation prior to 30 days after your receipt of the notice.

Termination of your rights under this section does not terminate the licenses of parties who have received copies or rights from you under this License. If your rights have been terminated and not permanently reinstated, receipt of a copy of some or all of the same material does not give you any rights to use it.

10. FUTURE REVISIONS OF THIS LICENSE

The Free Software Foundation may publish new, revised versions of the GNU Free Documentation License from time to time. Such new versions will be similar in spirit to the present version, but may differ in detail to address new problems or concerns. See http://www.gnu.org/copyleft/.

Each version of the License is given a distinguishing version number. If the Document specifies that a particular numbered version of this License "or any later version" applies to it, you have the option of following the terms and conditions either of that specified version or of any later version that has been published (not as a draft) by the Free Software Foundation. If the Document does not specify a version number of this License, you may choose any version ever published (not as a draft) by the Free Software Foundation. If the Document specifies that a proxy can decide which future versions of this License can be used, that proxy's public statement of acceptance of a version permanently authorizes you to choose that version for the Document.

11. RELICENSING

"Massive Multiauthor Collaboration Site" (or "MMC Site") means any World Wide Web server that publishes copyrightable works and also provides prominent facilities for anybody to edit those works. A public wiki that anybody can edit is an example of such a server. A "Massive Multiauthor Collaboration" (or "MMC") contained in the site means any set of copyrightable works thus published on the MMC site.

"CC-BY-SA" means the Creative Commons Attribution-Share Alike 3.0 license published by Creative Commons Corporation, a not-for-profit corporation with a principal place of business in San Francisco, California, as well as future copyleft versions of that license published by that same organization.

"Incorporate" means to publish or republish a Document, in whole or in part, as part of another Document.

An MMC is "eligible for relicensing" if it is licensed under this License, and if all works that were first published under this License somewhere other than this MMC, and subsequently incorporated in whole or in part into the MMC, (1) had no cover texts or invariant sections, and (2) were thus incorporated prior to November 1, 2008.

The operator of an MMC Site may republish an MMC contained in the site under CC-BY-SA on the same site at any time before August 1, 2009, provided the MMC is eligible for relicensing.

ADDENDUM: How to use this License for your documents

To use this License in a document you have written, include a copy of the License in the document and put the following copyright and license notices just after the title page:

```
Copyright (C)  year  your name.
Permission is granted to copy, distribute and/or modify this document
under the terms of the GNU Free Documentation License, Version 1.3
or any later version published by the Free Software Foundation;
with no Invariant Sections, no Front-Cover Texts, and no Back-Cover
Texts.  A copy of the license is included in the section entitled ''GNU
Free Documentation License''.
```

If you have Invariant Sections, Front-Cover Texts and Back-Cover Texts, replace the "with...Texts." line with this:

```
with the Invariant Sections being list their titles, with
the Front-Cover Texts being list, and with the Back-Cover Texts
being list.
```

If you have Invariant Sections without Cover Texts, or some other combination of the three, merge those two alternatives to suit the situation.

If your document contains nontrivial examples of program code, we recommend releasing these examples in parallel under your choice of free software license, such as the GNU General Public License, to permit their use in free software.

Index